Body Science:

The New 21st Century Understanding Of How Your Physiology Really Works, Leave The Myths And Lies Behind, Get Healthier Than You (Or Your Doctor) Ever Imagined And Avoid Chronic Disease

by Dave Champion

Body Science: The New 21st Century Understanding Of How Your Physiology Really Works, Leave The Myths And Lies Behind, Get Healthier Than You (Or Your Doctor) Ever Imagined And Avoid Chronic Disease

Copyright © 2019 by Dave Champion

All rights reserved. No part of this book may be used or reproduced by any means, graphic, electronic, or mechanical, including photocopying, recording, taping or by any information storage retrieval system without the written permission of the publisher except in the case of brief quotations embodied in critical articles or reviews.

Because of the dynamic nature of the Internet, any web addresses or links contained in this book may have changed since its publication and may no longer be valid.

ISBN: 978-0-578-59980-9
Publisher: Dave Champion

DEDICATION

To all those who have lost family members or friends to chronic disease without knowing or understanding they were taken from you years earlier than necessary had the world known what you are about the learn.

"Here's to the crazy ones. The misfits. The rebels. The troublemakers. The round pegs in the square holes. The ones who see things differently. They're not fond of rules. And they have no respect for the status quo. You can quote them, disagree with them, glorify or vilify them. About the only thing you can't do is ignore them. Because they change things. They push the human race forward. And while some may see them as the crazy ones, we see genius. Because the people who are crazy enough to think they can change the world, are the ones who do."

~ Rob Siltanen

DISCLAIMER:
I am not a medical doctor.
This book does not provide "medical advice".

This work is produced and dissiminated based on the author's unalienable right as a citizen of a state of the Union to speak publicly on any subject.

PUBLIC SPEAKING:
To arrange a speaking engagement
with Dave Champion, go to drreality.news/speak

CONFLICT OF INTEREST
DISCLOSURE STATEMENT:
Dave Champion has received no compensation of any kind, or in any form, from any person, entity, or organization in relation to the writing or publication of this work.

ADDITIONAL BOOKS
BY DAVE CHAMPION:
Income Tax: Shattering the Myths
Available at drreality.news

IN QUESTIONS OF SCIENCE,
THE AUTHORITY OF A
THOUSAND IS NOT WORTH
THE HUMBLE REASONING
OF A SINGLE INDIVIDUAL.

~ Galileo Galilei

TABLE OF CONTENTS:

INTRODUCTION
page 1

WELCOME TO BOOT CAMP
Chapter 1 - page 7

CAN WE TALK?
Chapter 2 - page 13

FIRST THINGS FIRST
Chapter 3 - page 21

LYMPHATIC LIPID SYSTEM
Chapter 4 - page 27

NUTRITIONAL ANTHROPOLOGY
Chapter 5 - page 35

HEPATIC LIPID SYSTEM
Chapter 6 - page 43

THOSE PESKY IMPLICATIONS
Chapter 7 - page 51

TURNING A BLIND EYE
Chapter 8 - page 59

NO SCIENCE NEEDED (FOR LIES)
Chapter 9 - page 67

THE TRUTH AIN'T PRETTY
Chapter10 - page 77

FACTS. THEY MATTER.
Chapter 11 - page 85

THIS CHANGES EVERYTHING!
Chapter 12 - page 117

A BRAVE NEW WORLD
Chapter 13 - page 131

INSULIN INTOLERANCE HYPOTHESIS
page 145

ACKNOWLEDGMENTS
page 151

BODY SCIENCE
INTRODUCTION

You don't know it yet, but you just sat down and strapped-in for one of the wildest rides you're ever going to take!

Before I launch you down the track, let me share with you how this book came to be.

Several years ago I was experiencing some seemingly minor health-related issues that were annoying me. I started searching for relevant factors in the hope of finding a solution.

That simple first step ended up propelling me into the study of physiology.

Surprisingly, I fell in love with the subject!

Not only did I fall in love with the science of physiology, it came easily to me.

Being a habitual learner, I have studied a number of subjects over the decades. None "clicked" for me the way physiology has.

How the body does what it does simply makes sense to me. It is logical. It is predictable (mostly). I enjoy every minute studying human physiology.

Eventually the time came to expand my studies from the numerous systems of the body and begin looking at how nutrition - what we put in our mouths - impacts the body's operation.

That moment was the genesis of this book.

Having studied the internal chemical and physical functioning of the body's various systems before I started reading data from the nutritional research industry, I was left scratching my head at the conclusions of "nutritional experts".

To believe, or agree with, many (if not most) of the conclusions from the nutritional research industry I would have had to set aside the science of how the body works.

Phrased more pointedly, either the science I'd studied concerning how the body functions was flawed, or the conclusions of the nutrition experts were flawed. Both could not be correct because they were at scientific odds with one another.

The desire to understand, and settle, this conundrum propelled me down the road of investigating the "facts" put out by the nutritional research community. I put "facts" in quotes because when I started investigating I had assumed statements put out by various nutritional research organizations (public and private) were proven science.

I was stunned (to say the least) to find that almost none of the information put out - the advice given to the American public - by any of the numerous 'authoritative sources' on diet or nutrition over the last 60 years was based on science, or established as factual via scientific testing.

I imagine the near-complete lack of scientific substantiation comes as a surprise to you as well.

I suspect the public assumes the various 'authoritative sources' would never tell the American people how, as an example, to significantly reduce the odds of getting heart disease unless the information was scientifically proven to be factual.

Can you imagine my shock when, as I dug deeper and deeper into the research, I discovered that virtually none of it was scientifically proven? It had become "public policy" and "medical industry dogma" without meaningful research substantiating it!

I was appalled at the thought of hundreds of millions of Americans altering their diets, believing they were preventing chronic disease, based on information they presumed in good faith was proven science, when in fact the information was nothing more than hypotheses. In other words, conjecture!

Even if there was evidence suggesting these hypotheses were correct, those issuing the "official position" still should have been transparent with the public. Whoever released the information to the public should have told the American people plainly, "These are not proven facts. This information represents theories we hope will be proven in the future. At this time it is conjecture."

How differently might the public have viewed the information, and responded to it, if those putting hypotheses into public discourse had been transparent and forthright?

But it's worse than that. A large percentage of the hypotheses were horribly flawed - and plainly so.

In many cases, the weight of other data (often intentionally excluded from consideration by the hypothesis creator) made the hypotheses dubious from day-1. Yet they were presented to the public as established scientific fact.

One might consider whether a lack of forthrightness - the intentional concealment of the truth about the fundamental nature of the information - might indicate a lack of integrity.

If that thought crossed your mind, not only are you correct, but it is unlikely as you read these words you yet grasp how bad that problem has been. You will discover the full scope of that problem in this book.

It may be difficult to accept at this point - here in the Introduction - that virtually all of the information you've been given over the last several decades about what will keep you healthy and prevent chronic disease has been thoroughly inaccurate.

Not only has it been inaccurate, but in many cases it has been 180 degrees opposite of the facts; opposite of the science.

Phrased more directly, while authority figures were telling you that if you followed their advice you'd be less likely to get chronic diseases, the science of the matter is that if you followed their advice you were more likely to get chronic diseases. Much more likely!

The cold hard truth is the dissemination of grossly inaccurate information to the American people about nutrition and its affects on human health is the primary engine that has driven America into the worst health crisis mankind has ever known; a greater percentage of the American public having chronic diseases today than any people, anywhere, at any time in history.

But please…do NOT believe me.

That is a theme in all my various works; "Please do not believe me." I encourage you to do your own in-depth research after you complete this book.

My goal in this book is to give you that 35,000 foot view so you can see the Big Picture.

My priority is to take you from A to Z on the subject of how various nutrients impact whether you will be healthy or get debilitating chronic diseases.

Next, I will open your eyes to the almost endless falsehoods with which you've been bombarded by 'authoritative sources' over the last 60 years. I am not speaking of trivial falsehoods. In reality, we are talking about health-devastating falsehoods.

I also want to keep the material compact. In today's hectic world being direct and succinct is helpful and allows more people to make the time-investment needed to get the facts.

Covering such a broad array of inter-connected issues in a succinct manner means I cannot drill down into the details of every topic. That is one of several reasons I encourage you to do your own research.

My style of informing is to begin by providing you the necessary foundational knowledge so that your journey, and arrival at the ultimate destination, is interesting, compelling, well-understood, and well-appreciated. That foundational information is apparent in the early chapters.

The origin of this book is my disgust and indignation (to put it mildly) with the non-science-based agenda-driven false information that has been presented to you as 'scientific fact' for the last several decades, and the devastating health consequences those falsehoods have wrought in hundreds of millions of lives. In many cases tragic and heartbreaking consequences.

What that means is, at its core, this book is about physiology. About your body; your health.

Yet, as you will discover, it is so much more than that.

You're already strapped in. Get ready to launch!

My hope is this book will be a life-changing revelation for you.

My sincerest desire is that when you complete the final page, and close the cover, you will say to yourself, "Holy cow! That was an astounding experience. **This changes everything."**

CHAPTER ONE

WELCOME TO BOOT CAMP

"It's easier to fool people than to convince them that they have been fooled."
~ Mark Twain

The goal of this book is to give you a completely different outlook on nutrition and health than you've ever had before. An outlook that will help you avoid the chronic diseases so many of our friends, neighbors and family members are getting, or already have.

It is an outlook that, if applied, will make you robustly healthy, happy, and free of chronic disease.

This book will give you that outlook…if you let it.

Allow me to hijack and alter a Ronald Reagan quote; It isn't so much that Americans are ignorant about health and nutrition. It's just that they know so many things that aren't so.

Is it possible to receive an entirely new paradigm, no less accept it, if one's mind is chock full of non-factual information he/she believes factual?

Since virtually everything American society has ever heard from "the authorities" about health and nutrition has been wrong, does it not stand to reason you likely have ideas about nutrition and health that are based on that bad information?

If that non-factual information was presented to you as factual, and you didn't know it was incorrect when you encountered it, why wouldn't you have adopted it as factual? Especially if everyone else in society believes it too!

With that possibility in mind, let's run through a quick exercise.

Do you know that no matter how many times you've been told something is true, if it's not factual the number of times it's repeated doesn't make it factual?

Do you know that no matter who/what told you something non-factual, you viewing that source as credible or authoritative did not transform non-factual into factual?

Do you know that if a panel of "experts" told you something non-factual, the level of respect afforded that panel by the public does not transform non-factual into factual?

Do you know that if the government told you something non-factual, the fact that it came from government does not transform non-factual into factual?

Do you know that if your doctor told you something non-factual, the fact that it came from your doctor does not transform the non-factual into factual?

Do you know that if organizations such as the American Heart Association, or the American Diabetes Association, tells you something non-factual, the fact that it came from a big-name organization does not transform non-factual into factual?

Do you know that if your trainer told you something non-factual, the quality of that relationship does not transform the non-factual into factual?

That slew of "Do you know..." questions may have felt redundant. You may have thought to yourself, "Of course none of those things make the non-factual, factual. That's so obvious. Why is Dave bothering with this?"

Remember you said that in just a moment when I list nutritional/dietary myths you have likely believed to to be factual your entire life!

Because you have believed these nutritional/dietary myths your entire life, when you read the list your gut-reaction is likely going to be that I'm full of it. But that will be an emotional response, right? It has to be because you don't actually know the science behind the nutritional myths you believe. A one or two sentence 'scientific' explanation you read in an article, or were told by (as an example) a personal trainer, does not constitute 'knowing the science'.

Yet science is the deciding factor, not what you've been told, right?

In many cases you assumed there was credible science behind what you were told, when in fact there never was, as you will see in the chapters ahead.

In some cases what you were told felt intuitively correct. Based on that you adopted it as a truth in your life. Yet we know having a 'sense' that something is correct also has no bearing on facts or science.

Consider this...

Research shows the things we were told earliest in our lives - the things we were told **first** - tend to be the most resilient when challenged.

In other words, if you were told at 14 years of age that a certain thing is true - and you accepted that as true and integrated it into your knowledge base at that time - the amount of evidence required to change your mind at 30 would be substantial. In fact, there may be no amount of evidence that can alter your impression of what is true!

If you were told something at 24 - and you accepted it as true and integrated it into your knowledge base - the amount of evidence required to change your mind at 40 would be substantial. And so on.

That phenomenon remains operative even if **zero** evidence was provided to you at 14, or 24 (or 50!). If you have believed it as a truth for many years, it is a tall order to unwind that belief later.

"Later" - such as...now.

Further, a non-factual "truth" is reinforced simply by acting on it.

Research shows that once you take **action** based on something you believe, you become significantly more convinced of its "rightness". Simply by acting upon something you believe to be true, its "rightness" is enhanced and reinforced in the mind.

The more times, or the longer duration of time, your actions rely on that belief, the more it becomes etched in your mind as a truth.

The human mind is fascinating, isn't it?

All of that occurs at the subconscious level and the vast majority of people are completely unaware of those dynamics.

Understanding how the above mentioned phenomena operates, let's look at a (partial) list of nutritional "facts" you've almost certainly heard, and likely accepted as "true" without viewing a shred of actual scientific evidence.

1. You must keep saturated fat intake low if you wish to avoid heart disease.
2. Vegetables are an essential part of a healthy diet.
3. Fruit is part of a healthy diet.
4. Fiber is an important dietary substance for health.
5. A vegetarian diet is good for preventing chronic disease, such as heart disease.
6. A high-carbohydrate diet is best for preventing heart disease.
7. Eating something from each of the major food groups daily is the best way to stay healthy.
8. Eating a "balanced diet" is the best.
9. Any fat you consume should be "healthy fats" such as polyunsaturated fats from vegetable oil.
10. Eating in moderation is the key to good health.
11. The way to lose weight is to make sure the number of calories you consume is less than the number of calories you expend.
12. High intake of red meat increases your odds of getting cancer.
13. Whole grain products are an essential part of a healthy diet.

I could go on and on, but I'm sure you've heard most, if not all, of these pervasive false narratives and believe at least some of them to be factual.

Not one of them is factual.

That I am telling you things you absolutely believe to be true, are not, may have you thinking you wasted your money buying this book.

But that is a knee-jerk emotional response, isn't it?

Can you honestly say your belief in any of the above statements is the result of you reading clinical study reports? I don't mean articles in which a reporter told you what is (allegedly) in a clinical study report.

Have you actually sat down and read reports from numerous clinical trials addressing the issues in the above list? How about even one?

For 99.9% of you the answer to that is 'no'. Yet for some reason - which does not include scientific evidence - you believe certain things to be true. Perhaps adamantly so.

At this point you may be thinking, "If all those things I thought were true, aren't, what can I believe; who can I trust?"

Congratulations! You got the point.

So…who can you trust?

I certainly do **not** want you to trust me!

I don't want you to trust anyone.

I want you to trust **evidence** and nothing else.

At this point allow me to apologize if this chapter rubbed you the wrong way a bit. I entitled it "Boot Camp" for a reason.

In military boot camp the goal is to tear down how a recruit thinks and acts and remake him/her into the military member the government wants.

In this chapter I wanted to do the first part; tear down, or at least illuminate, your previous behavioral pattern of believing - trusting blindly - that which you have not verified by examining the science.

Why was it important to put you through this short "boot camp"?

Moving forward we're going to discuss the physiology of your body - and a lot more.

If, at every turn, you were thinking "Wait! That doesn't comport itself with [fill in any unverified 'belief' here]", you'd have an exasperating time ahead. You also may not have been able to see the data objectively. Hopefully you will now.

Does that mean I want you to accept uncritically anything or everything in this book? Hell no! But I do want you to let go of your preconceived ideas/beliefs and "go with the flow" as we move forward.

When you've completed the book I encourage you to do your own research on its content.

With "Boot Camp" behind us, let's get rolling!

CHAPTER TWO

CAN WE TALK?

"The beginning of wisdom is to call things by their proper name."
~ Confucius

You shouldn't need to learn an entirely new language to understand how your body really works and how to stay healthy. Yet in reality that is somewhat unavoidable.

One of the things that slowed my initial progress in learning physiology was that everything has unfamiliar medical or science-specific names. I'm going to endeavor to make the necessary vocabulary here as simple and streamline as possible for you.

You'll see the phrase "For the purpose of this book…" repeated in a number of definitions below. What that means is the word/term is more involved than how I am defining it here, but the more expansive elements of the definition are not relevant to anything we'll be addressing. As an example, glucose is found in plants and algae. Since we won't be discussing plants and algae, why include that in the definition?

Given the scope of information I'm going to share with you, the list of terms is actually rather short.

DON'T LET THESE DEFINITIONS INTIMIDATE YOU!

When I use these terms in context later, everything will fall into place beautifully. I promise.

You might want to place some form of bookmark at this chapter now in order to quickly and easily return here if, at some point, you want to double-check the meaning of a term.

A quick note about…"NOTE".

When "NOTE" appears in the text, the information provided is not essential to understanding the topic, but provides additional/richer context. In chapters 3 through 13, NOTES are numbered sequentially, per chapter, enabling you to refer to a NOTE (as an example), "NOTE3 in chapter 12."

GLOSSARY OF TERMS:

Blood - It is 92% water. Transports oxygen and nutrients to the cells. Removes waste from the cells. (Glucose travels easily & freely in it. Lipids require a protective "shuttle" to get them through the blood to their destination.)

Vascular System - The system of arteries and veins through which blood circulates throughout the body.

Oxygen - An essential life-sustaining element. Every single one of the 100 trillion cells of your body require it - at all times. The need of every cell in the body for oxygen is one of the reasons we breath. Oxygen is transported (in the blood) from the lungs to every cell in the body.

Fatty Acid - The most fundamental component (chemically speaking, at the molecular level) of what we think of as "fat". Fatty acids cannot travel through the blood unless in a protective "shuttle bus", such as lipoprotein.

Lipid - For the purpose of this book, HDL, LDL, IDL, VLDL and cholesterol.

Lipid Panel - A test to determine the quantity of certain lipids in the blood. The test is performed by drawing samples of blood and having a laboratory run specific procedures on the samples. The results are provided in a format that allows a person viewing the report to determine, via a numeric scale, how much of various lipids were present in the blood at the time the blood sample was taken.

Cholesterol - A lipid. Essential for all animal life. A key substance in a number of essential functions in the human body, cell structure being the most significant. Incorrectly attributed (for decades) as a causative aspect of atherosclerosis. In reality, a heart-protective substance.

Triglycerides - A lipid that can be consumed in the form of dietary fat, as well as be synthesized in the liver from glucose.

Phospholipid - A form of lipid that is a major component of all cell membranes.

Lipoprotein - A group of specific proteins that combine with and transport lipids in the blood.

> **VLDL** - Very Low Density Lipoprotein
> **IDL** - Intermediate Density Lipoprotein
> **LDL** - Low Density Lipoprotein
> **HDL** - High Density Lipoprotein

Apolipoprotein - A specific type of protein that bind various lipids together to form lipoproteins. They transport lipids (and fat soluble vitamins) in the blood and lymph fluid. There are numerous variants of apolipoprotein. (Apolipoproteins are virtually always abbreviated.) In this book we will discussing certain specific apolipoproteins, such as ApoB100, ApoB48, ApoCII, ApoE, and a few others. **Don't let this definition intimidate you!** I'll lay them out clearly and easily for you when addressing the relevant issues as we move ahead.

Chylomicron - A lipoprotein particle created in the intestines that travels through the lymphatic and blood circulatory systems serving as a 'shipping container' and delivery system for triglycerides to reach the cells of your body.

Synthesize - To create a substance by converting another substance into the needed substance, often involving hundreds of complex chemical processes. Example: The liver synthesizes triglycerides (fatty acids) from blood glucose.

Sugar - For the purpose of this book, sugar means 'sucrose', which is plain white table sugar; the substance people have used for hundreds of years to sweeten their coffee or tea. Table sugar is essentially equal parts glucose and fructose. (Don't confuse sucrose with Sucralose. Sucralose is commercial non-sugar sweetener most commonly known by the trade name, Splenda™.)

Glucose - For the purpose of this book blood; 1) Blood sugar. 2) One half of the molecules that comprise what we know as "table sugar". Glucose travels freely through the blood.

Glycogen - A modified form of glucose stored in various tissues of the body, most notably the liver (small volume) and skeletal muscle tissue (larger volume).

Monosaccharide - The most basic chemical form of what we call "sugar". Any carbohydrate consumed, in any form, will be broken down to a monosaccharide in the intestines before it can move through the intestinal wall into the blood stream; then referred to as "glucose". A monosaccharide is the most basic unit of carbohydrates.

Fructose - A unique form of sugar found in honey, tree and vine fruits, flowers, berries, and most root vegetables. Unlike every other forms of sugar, when fructose is digested it moves directly from the intestines to the liver and does not raise blood sugar levels. It is the primary ingredient in the commercial product, "high fructose corn syrup".

Liver - Second largest organ in the human body. Performs more than 150 distinct functions. It is 100% indispensable to life. The entire body is - scientifically speaking - amazing. But the liver may very well be the most amazing in a constellation of amazing elements. Several of its most fundamental functions are the focus of parts of this book.

Pancreas - For the purpose of this book, it is the organ that produces the body's insulin.

Insulin - A hormone produced by beta cells of the pancreas. Its impact is greatly misunderstood by the medical community.

Atherosclerosis - A condition characterized by excessive deposits of fatty-material plaque within the artery walls. Commonly referred to by the public as "heart disease" (though heart disease involves more issues than just atherosclerosis).

Lymphatic System - A key part of the body's circulatory system. Connects to, and interacts with, the blood circulatory system. Moves plasma and has a significant role in the body's immune function.

Thoracic Duct - For the purpose of this book, it is the lymphatic system "pipe" through which chylomicrons enter the blood circulatory system.

Lymphatic Lipid System (LLS) - The primary method used by the body to distribute dietary fatty acids to the cells. LLS is one of only 2 "lipid systems" in the body, the other being the "Hepatic Lipid System" (HLS).

> **NOTE:** *"Lymphatic Lipid System" is not a recognized medical phrase. It needs to be. As we proceed you'll understand why being able to effectively communicate a distinction between the LLS and HLS is so critical to a proper understanding of how the body truly operates. How do we educate people - how do we draw critical distinctions for them - if meaningful descriptors for differing systems do not exist? I created the title of Lymphatic Lipid System to differentiate these critical systems.*

Hepatic Lipid System (HLS) - The process whereby the liver converts excess blood glucose to triglycerides, packages them up in VLDL particles and sends the VLDL particles into the blood to store the unneeded triglycerides; the most frequent/common place for storage being adipose tissue.

> **NOTE:** *"Hepatic Lipid System" is not a recognized medical phrase. It needs to be. As we proceed you'll understand why being able to effectively communicate a distinction between the LLS and HLS is so critical to a proper understanding of how the body truly operates. How do we educate people - how do we draw critical distinctions for them - if meaningful descriptors for differing systems do not exist? I created the title of Hepatic Lipid System to differentiate these critical systems.*

Glucosis - A state in which the 100 trillion cells of the body use predominantly glucose as their source of energy.

> **NOTE:** *There are only two "hemispheres" concerning which mode of operation the body uses to fuel the cells. One is glucosis, the other ketosis.*

> **NOTE1:** *Glucosis is not a recognized medical term. Oddly, the medical community has never attributed a name for the mode in which the body uses glucose as its primary fuel. I created the term "Glucosis" to correct that problem.*

> **NOTE2:** *The word "glucosis" has been used less than a dozen times in U.S. medical literature since 1854. It frequently appears to be a typographical error, and has been used to denote a liquid solution containing a high amount of glucose.*

> **NOTE3:** *Dr Robert Atkins (founder of the erstwhile Atkins Diet) used the word "glucosis" as a means of illustrating to his patients what takes place when they reduced carbohydrate intake.*

Ketosis - A state in which the 100 trillion cells of the body use ketones and triglycerides for energy rather than glucose.

> **NOTE:** *There are only two "hemispheres" concerning which mode of operation the body uses to fuel the cells. One is ketosis, the other glucosis.*

Mitochondria - For the purpose of this book, the 'machinery' (technically, organelle) within almost every cell in the human body that oxidizes glucose, fatty acids, and ketones to create energy.

Macro - Shorthand for each of the three major macronutrients found in the food we consume; protein, carbohydrates, and fat.

Standard American Diet (SAD) - Research shows that American men eat roughly 400 grams of carbohydrates each day; women 300-350. A majority of these carbs are in the form of highly processed/refined foods.

Medical Advice - Information, recommendations, guidelines, or instructions provided to you by a doctor licensed to pratice medicine and with whom you have a professional relationship. Also includes information, recommendations, guidelines, or instructions provided by any personnel working under the doctor's authority.

Now that we have a common understanding of the terms you'll be seeing, let's get to it!

CHAPTER THREE
FIRST THING FIRST

*""Order and simplification are the first
steps toward mastery of a subject."
~ Thomas Mann*

In the next chapter we will begin the journey into the knowledge that may very well save your life.

Before we begin I'd like to share a few points I ask you to keep in mind as we move forward.

EXPECTATIONS

To say the body's chemistry is complex would be perhaps one of the world's greatest understatements.

The body's chemistry is, in fact, a rabbit hole. No matter how deep you go, the hole goes deeper. Ultimately, if you go deep enough, you will reach a point where man's knowledge is simply no longer sufficient to answer the questions we find.

While this book is rooted in the science of physiology, it does not teach you that field of study.

People pursuing various professions spend years studying physiology in order to gain a command of the subject. Obviously this book cannot accomplish that.

I'm going to break down highly complex material in a way everyone can understand. I'm going to stay tightly focussed on just what is important for you to know.

In the previous chapter I shared some basic vocabulary so we can communicate effectively and you will understand what is presented. Understanding the information is essential to applying it successfully. The purpose of this book is to empower you to do exactly that.

As I mentioned in chapter 1, upon completion of this book I encourage you to do as much independent research as you're willing. I'd estimate that about 98% of the information in this book can be verified independently on-line without too much effort.

I want to set the expectation that while you will be "exposed to" physiology, teaching it is beyond the scope of this book.

EVERYTHING IN DUE COURSE

I would ask you not to skip ahead, even if you feel you are familiar with some of the material.

It is possible some readers have a background in physiology, biology, medicine, etc. and feel there is no need to read what they already know. I appreciate the sentiment but would nevertheless ask you to not jump ahead. Here's why…

Millions of people could attend colleges or universities and learn physiology without ever coming to the conclusions I lay out in this book.

The value of this book is not so much in the raw material presented, as it is **how** the material is presented; how it flows together…how it connects the dots. The order in which the information is presented - and the important details along the way - are fundamental to properly understanding the Big Picture.

I will at times veer away from physiology to make a particular point, and then return to it. There is a method to my madness! Sit back and enjoy the ride. You have my word everything will come into crystal clear focus as I bring the pieces together.

EXCLUSIONS

This book is about how to live a life free of chronic debilitating illnesses/conditions such as insulin resistance, Syndrome X (aka; metabolic disease), hypertension, type-2 diabetes, heart disease, cancer, Alzheimer's, and more.

That said, this is not an "everything book". It cannot address every variation or every question a reader may have. My goal is to make the material as streamline as possible. That will maximize your absorption and retention of the information, thus best preparing you to apply it successfully in your life.

Because the human body is immensely complex there are many permutations in how it functions based on various external and internal influences. In other words, the body has what you might think of as back-up systems it can employ when ordinary conditions are altered. In some instances it has back-ups to the back-up! This book addresses some of those. It cannot address them all.

If you wish to know more - independent research!

BIO-INDIVIDUALITY

Each of us is unique in many ways.

While your genome is identical to the rest of mankind in the key elements that define the human race, your genome varies from that of other people in countless smaller particulars.

Those distinctions are often referred to as "bio-individuality".

Let's look at a few examples.

Males who smoke heavily have a 61% greater chance of getting lung cancer (over time) than males who have never smoked a cigarette. Yet 75.6% of male heavy smokers die without getting lung cancer. What differentiates the 24.4% from the 75.6%? Very likely genetic characteristics, i.e. bio-individuality

Some people drink a fifth of distilled spirits every day and live well into old age (though not many). Most get cirrhosis of the liver and die well before old age. The difference? Bio-individuality.

If two people get the same tattoo by the same artist on the same day, one may heal completely in 12 days, the other in 23 days. All other factors being equal…bio-individuality.

As with anything, the principle of bio-individuality can be misapplied. People being…well…people, some will pursue personal preferences that have no actual basis in physiology and justify it as "bio-individuality".

THE EXCEPTION DOES NOT DISPROVE THE RULE

We live in a time when people seem to enjoy being contentious. Social media seems to have enabled this to an extreme I've not seen earlier in my life.

One tactic of those who wish to be contentious is to offer an 'exception' to (supposedly) undermine the rule. Phrased another way, they attempt to argue that because something is not aways true, it is not generally true.

Need I point out the logical fallacy?

Let me give you an example. If I were to state that cancer strikes older people in much larger numbers than young people, the contentious person might say, "Then how do you explain childhood cancer?"

The contentious person is attempting to posit that since a small percentage of children get cancer that invalidates the more general - and factual - observation that the vast majority of people who get cancer are older.

Here are the facts:
- 36% of those diagnosed with cancer are 75 or older and another
- 53% are between 50 and 74 years of age. In other words, 89% of all people who develop cancer are 50 or over.

It is easy to see how this type of 'challenge' from a contentious person is counterproductive to getting at the truth.

GENETIC DISORDERS

The data and conclusions in the this book are for those whose bodies are functioning free of genetic disorders. This book expressly does not address genetic disorders.

There are two reason for this.

First, genetic disorders is a very specialized area of study/knowledge. If I tried to address that subject this would not be a book; it would be a multi-volume set of books.

Second, the percentage of people who have genetic disorders affecting things like liver function, pancreatic function, cholesterol production, ApoE production, etc., is small. I do not mean to be disrespectful to, or dismissive of, people who have such conditions. There is simply no way to achieve the objective of this book and address those very specialized needs.

Every statement in this book should be read to exclude physiological realities that exist due to genetic abnormality.

CAUTIONARY NOTE: *If you have a genetic disorder that involves anything discussed in this book you should ABSOLUTELY NOT act on the information in this book unless or until you have discussed it with your medical practitioner or specialist.*

CHAPTER FOUR

LYMPHATIC LIPID SYSTEM

"We celebrate our ability to create machines that move as man, yet we take for granted the miracle that is the human body."
~ David Fearnhead

Welcome to the beginning of the story! This is where the rubber starts meeting the road.

For those who are not immediately gripped by curiosity of such matters, this chapter may seem a bit "sciencey".

The reason it might seem that way is because you are not yet able to see the Big Picture from the 35,000 foot level. If you could, you'd look down at this chapter and say, "Ah ha! I would not have been able to understand the healthy functioning of the body versus unhealthy functioning of the body without chapter 4 under my belt!"

Also, I assure you the Lymphatic Lipid System - this thing of which you've not before heard - is significant. Incredibly so!

A good way to look at this is as if you are a detective working to solve a case! The allegation is the suspect manipulated a database and by doing so ripped people off. If you can prove he manipulated the database, you've got him! But in order to prove he manipulated it, you have to first understand how the database operates properly. Only with an understanding of how it properly operates can you then determine how the crime was committed!

In terms of systems of the body that deliver triglycerides to the cells, there are only two. The Lymphatic Lipid System (LLS) and the Hepatic Lipid System (HLS).

Let's start at the beginning. What is it that kicks the LLS into action? Answer: Eating something containing fat.

NOTE1: *For the purpose of highlighting the LLS we're going to ignore protein & carbohydrates. They use an entirely different pathway and hence are irrelevant to the LLS.*

Of the 3 macronutrients - protein, carbs & fat - only fat does not "digest" in your intestines.

Fat is handled differently. It is emulsified.

To keep it simple, that means dietary fat in the intestines is broken down to its most basic molecular structure; individual fatty acids. For convenience, we can think of these fatty acids as microscopic droplets of fat.

But here's the thing; fatty acids cannot move through the intestinal wall into the body unassisted. Additionally, even after moving through the intestinal wall, they cannot travel through the circulatory system unassisted.

NOTE2: *Fats/Lipids are "hydrophobic", meaning they cannot come into contact with water. The blood is 92% water. Similar for lymph fluid.*

In order for fatty acids to pass from the intestines, through the intestinal wall, thus entering the body, they must be bundled into something called a chylomicron. Being bundled into a chylomicron is also the assistance needed for the fats to transit the circulatory system.

A chylomicron is a combination of triglycerides, cholesterol, phospholipids, and apolipoprotein B48 - known as "ApoB48" (for structure). Only when the fats/lipids (triglycerides, cholesterol, phospholipids) are packaged up as a chylomicron can they cross through the intestinal barrier, thus entering the body.

NOTE3: *The intestines are not, medical speaking, considered "internal" to the body. The way the medical profession sees the intestines is merely a tube with openings at the top and bottom. While this tube runs through the trunk of the body, it is not considered an "inside" part of the body. Phrased another way, if you can put something in the top and it comes out the bottom, it is not an "internal" part of the body, medically speaking. I mention this so you understand why, as long as a substance remains in the intestines, it is not medically considered to be "inside" the body.*

Having passed through the intestinal wall, and thus now being "in" the body, what happens next?

The chylomicron enters the lymphatic system. No other macro takes this path. Only lipids entering the body through the intestines - dietary fat - enter the lymphatic system.

At this point the chylomicron is called a "nascent chylomicron". The nascent chylomicron travels along, doing nothing in particular, merely being carried through the lymphatic system by lymph fluid.

Eventually our chylomicron (actually, many of them) reach the thoracic duct, which empties its fluid into the vascular system, just above the heart. In other words, the fluid of the lymphatic system (called "chyle" at this point), containing the chylomicrons, exits the lymphatic system and enters the blood stream.

Our chylomicron is now traveling in the blood.

From this point forward our chylomicrons will remain in the blood until coming to the end of their existence.

As chylomicrons flow through the vascular system, they cross paths with HDL particles. (HDL particles are created in the liver). The HDL particles travel exclusively in the blood.

NOTE4: *HDL is often referred to as "good" cholesterol. In point of fact, it is neither good nor bad. It has a job to do like every other substance the body creates.*

When the chylomicron and HDL come in contact, the HDL transfers two specific apolipoproteins to the chylomicron; ApoE and ApoC2. After the transfer the HDL goes on its merry way.

If we think of a chylomicron as a "delivery vessel" (which it absolutely is), when it is "nascent" it cannot off-load its cargo. Absent getting ApoC2 from an HDL particle its cargo doors are locked shut. That's why it is referred to as "nascent" at that time.

Once it receives ApoC2 from the HDL, the cargo doors magically unlock! (OK…not really magic. Chemistry. But still pretty amazing.)

The chylomicron is now called a "mature chylomicron" and it gets busy delivering its triglyceride cargo to cells that need it. This is done by the chylomicron "docking" with cells throughout the body, off-loading its cargo, undocking, and moving on.

When the 'cargo hold' is empty the chylomicron once again links up with a passing HDL particle. This time it transfers the ApoC2 back the HDL particle. The cargo doors are now locked shut again and the "Closed for Business" sign is hung out! At this point it is referred to as a "remnant chylomicron".

Let me pause here for a moment to say that what I just described for you is what you need to know about this process in order to put everything together properly as we proceed. I have left out about…oh…say… a couple of thousand chemistry-based details on how what I described gets accomplished.

Not only would this be a very long dry chapter if I went into detail on the thousands of chemical processes involved, but more importantly, none of that is relevant to where you are heading in this story. (As always, if you desire to know the fine detail beneath the description I provide, I encourage you to research.)

Once the ApoC2 has been transferred back to an HDL particle, and the chylomicron is then a "remnant", it has only one thing left to do; make its way to the liver where it will cease to exist.

The liver recognizes a chylomicron containing ApoE and ApoB48 (but without ApoC2) as a chylomicron that needs to be broken down and its constituent elements recycled or removed from the body.

The presence of ApoE allows the chylomicron to "dock" within the liver on a LDL receptor cell. From there the liver may begin the process of breaking down the remnant and recycling its constituent elements. That process is known as endocytosis. The liver does a lot of that.

The liver may also excrete the remnant into the intestines via bile, where it will leave the body in your stool.

At this point we've tracked a chylomicron from birth to death.

Let's do a quick recap:

In the intestines dietary fat is emulsified into microscopic droplets of fatty acids.

Those fatty acids are packaged together with a structural protein into a "chylomicron".

Only when packaged as a chylomicron can the fatty acids cross the intestinal wall, entering the body.

Immediately after passing through the intestinal wall chylomicrons enter the lymphatic system.

Chylomicrons travel through the lymphatic system in their "nascent" form.

The lymphatic system dumps the chylomicrons into the blood stream via the thoracic duct.

Chylomicrons are provided specific proteins from HDL, thus changing the chylomicrons to "mature".

Mature chylomicrons travel through the blood off-loading triglycerides to cells that need it for energy.

When all the 'cargo' has been off-loaded, the chylomicron returns the ApoC2 protein to a passing HDL, thus becoming 'remnant'.

The remnant arrives at the liver, links up with an LDL receptor cell and is cannibalized for recycling or is excreted. The chylomicron exists no more.

I told you this would be a breeze!

KEY TAKE-AWAYS CONCERNING THE LLS

There are three key takeaways from this chapter.

Number One: Fat Remains Fat

What I am about to say may seem like I am needlessly stating the obvious, but it is important to state this reality plainly here and now so I can more clearly draw your attention (when the time comes) to a contrast you will need to recognize.

The statement is this: **In the Lymphatic Lipid System fat remains fat 100% of the time.**

Fat enters the body as dietary fat. It is emulsified into fatty acid droplets. The triglycerides (cargo) in the chylomicrons remains fatty acids. (Actually, 3 fatty acid molecules bonded to a glycerol molecule.) They remain fatty acids throughout their transportation to the cells. They are fatty acids when off-loaded to the cells. And they are fatty acids until the instant they are oxidized (burned) for energy by the cells of the body.

Stated succinctly, from the time you put the food containing fat in your mouth, until the moment the triglycerides cease to exist by conversion to cellular energy, they remain, at all times fatty acids.

Number Two: No Storage

The triglycerides transported and delivered via the LLS are virtually never deposited in the cells of your adipose tissue (belly fat) or visceral cells (fatty tissue between your internal organs). In other words, triglycerides delivered via the LLS are burned, not stored.

Number Three: A Constant

You will recall these definitions from Chapter 2...

Glucosis - A state in which the 100 trillion cells of the body use predominantly glucose as their source of energy.

Ketosis - A state in which the 100 trillion cells of the body use ketones and fatty acids for energy rather than glucose.

The LLS operates the exact same way in either mode of operation. What you just read happens precisely the same way in the bodies of people in glucosis and those whose bodies are in ketosis.

CHAPTER FIVE

NUTRITIONAL ANTHROPOLOGY

"Studying primitives enables us to see ourselves better."
~ Clyde Kluckhohn (Anthropologist)

The macro theme of this book rests on the premise that our genome - the genetic coding of our bodies - dictates our body's physiological "norm"; it's proper and healthy functioning.

The way I think of that "norm" is this...

If the body had a way to verbalize to us (without involving our conscious mind) it would say, "**This** is the factual and proper way I function. And **this** is how you should be fueling me to support that proper functioning."

If one does not believe that man's physiology, and what the Earth natively provided (pre-technology), operates in symbiosis, then the reasoning undergirding the premise of this book will be missing for that person.

Whether one believes evolution is the cause of our genetic coding - or a deity created it - or some wild & crazy space aliens did - or if one says, "Who the hell knows?" - your genetic coding exists. It has remained fundamentally unchanged for millions of years.

The earliest fossil record of humankind is 66 million years old. How much earlier humankind may have existed is unknown. 66 million years is considerably more than adequate to illustrate the premise upon which this book's conclusions are founded.

There are...broadly speaking...two models concerning man's existence.

One is that a creator created man, the Earth, and...well...everything. In this scenario, man has existed "as is" from day-1 and his genetic coding was written by his Creator.

The other model is that mankind, as we see him standing before us today, is the result of billions of years of evolution. In this model man's genetic coding is the result of (at a minimum) hundreds of millions of years of refinement and has remained unchanged for (at a minimum) hundreds of thousands of years.

The good news is, for the purpose of this chapter - and indeed, this book - either model works!

If we accept that our genetic coding dictates our body's physiological norm, then it is reasonable and logical to look back as far as we can with the intent of determining what the symbiotic relationship between Earth and man looked like before technology altered the relationship.

CLIMATE

Oddly, scientists claim to be fairly certain of what Earth temperatures were billions of years ago, or hundreds of millions of years ago, but much less certain about the era in which we know humankind existed. (It's not really "odd". It's simply easy to say one "knows" when there is no way to prove or disprove what is being asserted.)

If there are reliable numbers for more modern atmospheric conditions, such as...say...the last 500,000 years, they are difficult to come by. It seems that for most of that period the Earth was in an ice age. Depending on which source one looks at, this ice age started a couple of million years ago, and we're still in it. However, within this long-term ice age are periods referred to as "glacial" and "interglacial" during which temperatures became cooler or warmer, respectively.

Why do we care? Temperature would affect food availability, yes?

In the 20th century we knew the Inuit people, living about as close to the Arctic Circle as man can, had a significantly different diet than people living in a tropical rainforest on the Equator. Common sense stuff.

So then...what was man eating from...say...50 million years ago... until roughly 12,000 years ago, when the First Agricultural Revolution occurred?

NOTE1: *The First Agricultural Revolution was the beginning of the 'technology' I mentioned earlier that began altering the relationship that had existed between man and Earth for (at a minimum) tens of millions of years.*

I posit that question because I see that timeframe as reasonable to consider what our ancestors ate. If we agree that man and Earth enjoyed a symbiotic relationship for tens of millions of years, then what the Earth made available to man remains man's genetic coding today.

We need to take a moment here and acknowledge that what we are about explore is a "best guess" based on hypotheses built from precious few known/provable facts. That said, I believe biology, our knowledge of agriculture, and a dose of common sense can take us a good way down a rational road.

Did the Earth have 'seasons' each year during the last 50 million years? Of course. Seasons are a result of the tilt of the Earth's rotational axis during our planet's 365 day trip around the Sun. That has been a constant (in terms of billions of years).

Since there were seasons, there would be a time when plants are edible (in whole or in part) and trees would bear fruit; what we today call "harvest season". There would be a much longer period of the year when they would not. No different than today.

Further, we know from plant anthropologists that most edible plants we enjoy today looked and tasted nothing like what we are familiar with. Many items didn't exist, being the result of cross-breeding since the First Agricultural Revolution.

Edible plants that existed were smaller, less appealing (as we see it from today's perspective), with no known, fixed, location. Remember, 60,000 years ago (as an example) there were no farms, farmers, planted fields, or orchards. Even in Springtime, when edible plants were ripening, humans essentially had to stumble upon them by chance, or recall from memory where to find them from last year (presuming the person was in the same locale a year later).

Also, because these plants are edible to many other species, man had constant competition from a long list of mammals, reptiles and insects during the short period these plants were edible.

Precisely what volume of edible plants were available to our ancient ancestors we may never know. But we can be certain it was an incredibly small percentage of their overall food intake in the best of times (springtime and part of summer), and virtually nonexistent most of the year.

We should also consider that fruits and vegetables do not deliver as much energy as does animal flesh. This is because animal flesh is partially fatty tissue. Fatty tissue has double the calories per gram as carbohydrates, which is the predominant macro in fruits and vegetables.

Ancient man is often portrayed in our media culture as little more than a grunting creature, only marginally more advanced than a chimpanzee. The physiological reality is ancient man (at least as far back as we can credibly speculate) had much the same cognitive abilities as me and you. (Current thinking of the "experts" is ancient man's intellect matured more slowly than ours. That supposition is based on questionable interpretations from very limited data.)

Ancient man couldn't tell you what 2+2 equals [because he hadn't been taught, not because he couldn't be taught], but he was certainly well aware of what type of food satiated his hunger best, provided him the greatest level of energy, and stayed with him the longest time, thus allowing more rest before having to strike out for food again.

Ancient man's #1 job was survival. As such, animal flesh, with its far higher calorie (energy) count, would have been his preferred food at all times.

Earlier I mentioned glacial and interglacial periods. The glacial periods, when the Earth's temperature dropped, are measured in tens of thousands of years. The interglacial periods, when the Earth would warm, are measured in only a few thousands of years. In other words, the available data tells us that for a majority of the time period we're considering, temperatures were colder, not warmer.

This buttresses the view that edible plants and fruit were uncommon. Not many edible plants prefer colder temps. Colder overall temps also translate into a shorter season of availability, i.e., shorter Spring & Summer. Phrased another way, colder temps would have reduced the types of edible plants available, and the shortened "harvest season", constricting availability.

On the other hand, colder temps would not adversely affect the presence or quantity of game animals.

Though we can't know for certain, there are anthropological reasons to believe that during a time when few humans populated the Earth game animals were plentiful.

When we combine the lack of edible fruits and vegetables, both as a percentage of time available during the year and being uncommon to come across, with the fact that game was prevalent and early man would have preferred animal flesh whenever possible, there is little doubt what constituted the bulk of his diet.

NOTE2: *There are some folks who present false narratives about what ancient man ate. Their false narratives are usually driven by a modern agenda. Common agendas include things like "saving the planet" and/or preventing what they view as animal cruelty. However one may feel about such issues, supporting such agendas does not excuse promoting falsehoods.*

NOTE3: *A major U.S. news network recently ran a story intended to convince Americans that ancient man ate a high-carb diet. Their method of scamming readers was to use the diet of a current tribe in the Amazon Rain Forest that eats (mostly) as they presumably did for hundreds of thousands of years. The problem is - at least for critical thinkers - the Amazon Rain Forest is an eco-system vastly* *different from the rest of the planet. The diet of Rain Forest occupants would have been considerably different than the diet of ancient man anywhere else on the planet.*

NOTE4: *Some folks claim man only "learned to hunt" 25,000 years ago, or so. That is fabricated, agendized, nonsense. Man is a hunter by virtue of being, as my friend Robert Bub phrases it, a "non-obligatory carnivore". In other words, while man is technically an omnivore, by combination of preference and availability, his daily existence was overwhelmingly that of a carnivore. All meat-eaters are born with the instinct to hunt and are taught how to effectively do so by a parent.*

NOTE5: *Certain anthropologists like to say, "The diet of all ancient men was not identical. There was not just one diet." While that is certainly true, it is also irrelevant. Not everyone owns a cell phone today. Yet if we were discussing the 21st century what would be the point of saying, "Not everyone owned a cell phone?" The overwhelming majority of people do. Likewise, while not all ancient men ate the same diet, their preference for meat was the same. Ancient man on the savanna of Africa would desire to kill and eat a gazelle, while ancient man in what is today the State of Tennessee would desire to kill and eat a deer. Would the scarcity of fruits and vegetables be slightly greater or lesser in one locale than another? Of course. But a slight difference in the level of scarcity is insignificant.*

Understanding that ancient man ate predominately animal flesh, it is time to discuss a term that appeared in the definitions in chapter 2.

Ketosis.

Animal flesh contains only two macros - protein and fat; no carbohydrates.

With no carbs (or extremely low carbs) in one's diet the body operates in ketosis; a state in which the 100 trillion cells of the body use ketones (synthesized in the liver from fatty acids) and triglycerides (provided by chylomicrons we've discussed) for energy.

I should point out that while much of what we've discussed in this chapter are conclusions based on a combination of known facts, rational presumptions, and a dose of supposition, the human body operating in ketosis when on a no-carb or very-low-carb diet is a concrete physiological fact.

Although I will address ketosis and related issues more comprehensively in a future chapter, I will here draw your attention to these facts:

1. Man's diet was unchanged for millions of years.
2. That diet was predominately animal flesh.
3. Because animal flesh has no carbohydrates, the result was a state of ketosis.
4. Accordingly, man's condition for millions of years was ketosis.
5. Man's core genetic coding, if alterable over time, would only change across hundreds of thousand of years, best case. Perhaps millions.
6. The First Agricultural Revolution allowed a small percentage of mankind to begin consuming marginally more carbs, growing over time.
7. The First Agricultural Revolution occurred 12,000 years ago.
8. Zero alteration of core genetic coding occurs in 12,000 years, or anything close to it.
9. Accordingly, man's core genetic coding was - and remains - a state of ketosis.

CHAPTER SIX

HEPATIC LIPID SYSTEM

"The liver is the mother of the blood."
~ Jakob Bohme

"Hepatic" means 'having to do with the liver'.

There are only two 'systems' in the body that distribute new triglycerides.

In chapter 4 we looked at the Lymphatic Lipid System (LLS). Now we look at the Hepatic Lipid System (HLS). The common thread is both deliver "energy".

Allow me to explain.

Triglycerides = Energy. From this moment forward, whether reading it in this book, or wherever else you encounter the word "triglyceride", you should immediately think "ENERGY!" As far as the body is concerned, triglycerides are energy.

Further - and something you should keep in mind as you read on - whether we are considering the LLS, or the HLS, the focus of our inquiry is on the packaging and delivery of triglycerides (ENERGY!).

You will recall I mentioned earlier that the liver performs more than 150 distinct functions. Our discussion of the HLS does not include those. We are here looking exclusively at the synthesis and output of triglycerides by the liver, and what takes place in our bodies when the liver puts that energy (triglycerides) into the blood.

NOTE1: *In this chapter we will not be addressing what the liver does with fatty acids when the body is fasting. That is a different mechanism than the output of triglycerides synthesized from glucose, which is the HLS.*

When discussing the LLS, the relevant macro was fat. The discussion began with consuming dietary fat. We ignored the other 2 macros - carbs and protein - because they have no role in the LLS.

Now, with the HLS, we ignore dietary protein and fat, focussing exclusively on carbohydrates we ingest. We exclude dietary protein and fat because they have no role in the HLS.

All carbohydrates - whether table sugar or broccoli - are broken down in the intestines to monosaccharides. That is an absolute. There is no exception.

It's a simple, singular, and linear process; carbs > monosaccharides.

Monosaccharides are the simplest form of what we commonly call 'sugar'. When a sugar cannot be broken down (hydrolyzed) further, it is then a monosaccharide. Phrased another way, A monosaccharide is the most basic unit of carbohydrates.

Once carbs have been rendered into their most basic unit - monosaccharides - those monosaccharides pass through the intestinal wall directly into the blood. At this point the medical industry stops referring to them as monosaccharides and begins referring to them as "blood glucose". One moment they're monosaccharides, the next blood glucose.

The quantity of carbs consumed is the determining factor of how much glucose is present in the blood after eating.

Phrased another way, if I sit down and eat an ounce of broccoli there will be far fewer monosaccharides flowing into the blood as glucose than if I ate a bowl of pasta.

The next question we might reasonably ask is, what does the body do with that glucose?

A normal healthy body has a stable glucose 'baseline' when not consuming dietary carbs. In other words, if you didn't eat anything that added glucose to the blood, there remains a modest amount of glucose present in the blood at all times. That is healthy, proper, and essential. The body has several mechanisms by which to ensure that baseline remains stable in the absence of carb intake.

So...having not eaten anything in 6, 8, 10 or 12 hours while sleeping, you start the day with your glucose at baseline.

What happens when we consume dietary carbs? Answer: Blood glucose increases from baseline. In fact, once a person begins eating for the day, the baseline is out the window until the next morning, after a night of not eating. (This presumes like most people you eat multiple times a day. This is why you are asked to fast for at least 8 hours before a blood glucose test.)

How much does your blood glucose level elevate when you eat? That depends on how many carbs you consume. If you eat a modest amount of carbs during a meal your blood glucose level will rise a modest amount. If you eat a lot of carbs your blood glucose will increase a lot. (We discuss the speed of increase, and its significance, in a later chapter.)

Something has to happen to the elevated blood glucose because glucose left elevated is not healthy for organs and tissue.

When blood glucose levels rise the body responds by the pancreas producing insulin and releasing it into the bloodstream.

Most people are uncertain exactly what insulin does. Many folks know insulin lowers blood glucose, but most are unaware of how, exactly, insulin does it.

In a healthy person insulin production is scaled in relation to the amount of glucose elevation. If glucose elevates a modest amount, the pancreas produces a modest amount of insulin. If glucose is elevated a lot, the pancreas will product a lot of insulin. In other words, the pancreas produces enough insulin to address the situation created by the amount of carb-containing food you ate.

When insulin is produced and released into the blood, what does it do?

Insulin performs a number of functions, but for our purposes it does two things: (These two are the most significant things insulin does.)

1. It commands the cells to open up and receive nutrients traveling in the blood.
2. It signals the liver to use glucose as the substance from which to synthesize triglycerides.

A carbohydrate rich meal will produce high blood glucose. The high blood glucose level signals the pancreas to put insulin into the blood. In turn, the insulin signals the cells to open up and accept nutrients. Since the blood glucose level is high, the cells will take in as much glucose as they can accept. A significant qualifier is the cells take only as much as they can accept, and no more. (Skeletal muscle will absorb a minor amount of glucose [stored as glycogen], if needed.)

On a high carbohydrate diet (as the vast majority of Americans eat) after the cells have consumed as much glucose as they can, there will still be excess glucose in the blood. It can't stay there. That's harmful to organs and tissue.

The body's agenda is to always move the blood glucose level back toward baseline. When glucose is elevated something must be done in pursuance of that requirement.

Queue the liver synthesis of excess blood glucose into triglycerides.

The primary feature of the Hepatic Lipid System is the liver's ability to convert excess blood sugar into triglycerides, and under what circumstances it does so.

As mentioned a moment ago, elevated insulin signals the liver to synthesize the excess blood glucose into triglycerides.

The liver cannot store the triglycerides it creates. Stated plainly, as the liver synthesizes triglycerides it has to get rid of them. It does that by bundling them into lipoproteins known as VLDL; Very Low Density Lipoprotein.

These VLDL particles are structurally similar to the chylomicrons we discussed in Chapter 4. The most apparent distinction between chylomicrons and VLDL particles is that instead of having an ApoB48 protein for structure, as do chylomicrons, VLDL particles have an ApoB100 protein. (ApoB48 is only made in the intestines; ApoB100 only in the liver.)

If the amount of excess glucose in the blood is modest, then a modest amount of synthesis occurs and a modest number of VLDL particles set sail from the liver. If the level of excess glucose is high, then a lot of VLDL particles set sail.

Pop quiz time! What did I tell you to think of every time you see the word "triglycerides"?

ENERGY, right?

Probably 99.5% of Americans live in a state of glucosis [cells burning predominantly glucose for energy], so when their cells uptake a lot of glucose (as commanded by insulin) the role of triglycerides as a source of cellular energy is very limited. That limited need is met by the chylomicrons delivering triglycerides through the LLS, which does not require insulin to signal the cells to take in the fatty acids they need.

NOTE2: *Remember, chylomicrons deliver fatty acids from what you eat. VLDL deliver fatty acids synthesized from excess glucose. That glucose-to-triglyceride process being insulin-driven.)*

So...the cells have all the energy they need in the form of glucose; their predominant fuel when in glucosis. Whatever small need cells have for triglycerides has been more than adequately met by the chylomicron deliveries. Yet now the liver is pumping out VLDL particles filled with even more energy; energy for which the cells have no use!

Something has to be done with the unneeded energy now circulating in the blood.

Complicating matters is that the VLDL particles carrying that unneeded energy cannot return to the liver with the triglycerides still on-board.

If the triglycerides can't be oxidized in a cell for their energy capacity because the cells are already flush with energy resources, and they cannot stay in the VLDL particles, they obviously have to be put somewhere.

Your body's mechanism to resolve that situation is to **store** them.

The human body stores triglycerides it can't use in adipose cells; your fat cells.

How many triglycerides that cannot be used get stored in the adipose fat cells? All of them.

How many is "All of them?"

That's predicated on how much excess glucose is present in the blood. The pancreas will continue putting out insulin - which signals the liver to continue the conversion of glucose into triglycerides - until the blood has been cleared of excess glucose and returned to somewhere near baseline.

Let's do a quick review.

You eat.

If you're like most Americans, your meals contain a lot of carbohydrates.

Those carbs are converted to monosaccharides [the most basic form of sugar] in the intestines and pass into the blood, where they are then called 'glucose'.

When the body senses high blood glucose the pancreas puts insulin into the blood.

Insulin signals the cells to 'open up' and take in nutrients that are circulating in the blood.

The cells uptake as much glucose from the blood as they can, but no more.

Whatever glucose is left in the blood, which is usually a significant amount, is then synthesized from glucose into triglycerides by the liver.

Because the liver can't retain the newly manufactured triglycerides it puts them into the blood stream, packaged up as VLDL particles.

The VLDL particles **must** get rid of the excess energy they are carrying. They cannot return to the liver with the triglycerides still on-board.

The most common place for the fatty acids to be deposited (stored) is in your adipose tissue.

That is the HLS in a nutshell. Not too difficult to understand, right?

While the HLS operates in exactly the same way in the bodies of people living in glucosis or ketosis, it should be noted that our ancestors - living as they did in ketosis - would very rarely have had access to, or have eaten, sufficient carbs to kick the HLS into action. Phrased another way, for humans who lived in a state of ketosis for millions of years, while the HLS was present, it was for the most part dormant; something kept in reserve for emergencies.

Only rarely did our ancient ancestors encounter sufficient carb-rich food to drive their blood glucose level high enough that the HLS had to kick in and save the day.

CHAPTER SEVEN

THOSE PESKY IMPLICATIONS

"We are being conditioned, as a population…to accept thoughtless constant consumption as the new norm."
~ Jenna Wortham

At this point we have discussed the Lymphatic Lipid System (LLS), key elements of nutritional anthropology, and the Hepatic Lipid System (HLS).

It is now time to consider the implications to which those discussions lead us.

Let's begin our examination of the implications by taking particular note of the fact that triglycerides delivered by chylomicrons (LLS) do not require an increase in insulin from baseline to distribute their 'cargo' to the cells.

Phrased more pointedly, dietary fat eaten without accompanying carbs or protein produces no insulin response.

Why?

Because the distribution and cellular uptake of triglycerides (fatty acids) from food consumption does not require any increase in insulin.

You will recall that after off-loading their cargo of fatty acids to the cells, chylomicrons return the ApoC2 to a passing HDL while retaining their ApoE so they can readily attach to LDL receptor cells in the liver where they will either be consumed and recycled, or removed from the body via bile excretion.

Just as with LDL particles, chylomicrons cannot be processed by the liver until they have off-loaded their triglyceride cargo to the cells.

The lifespan of a chylomicron is 30 to 60 minutes. In other words, a chylomicron is formed in the intestines, travels through the lymphatic system, enters the blood via the thoracic duct, travels throughout the body distributing its cargo of energy to the cells, ending up in the liver for termination - all in 60 minutes or less.

That entire process occurs while your insulin level remains at baseline.

This is entirely different than the HLS, which requires elevated insulin as the signal to the liver there is a high blood glucose problem and the liver should immediately begin converting the excess blood glucose to triglycerides for storage in the body's fat cells.

Allow me to draw your attention directly to the distinctions of which I am asking you to take note.

1. Energy distributed via the LLS requires no increase in insulin above baseline and the chylomicrons' cargo being distributed to the cells is not stored; it is promptly oxidized for energy.
2. Energy distributed via the HLS does require elevated insulin. Not only to signal the liver to synthesize the excess blood glucose into triglycerides, but also to command the adipose cells to accept the liver-manufactured fatty acids (which are not needed for oxidization by even one cell in the body at that time).

The pervasive narrative from the nutritional research and medical communities is (and has been for decades) that the storage of unneeded energy in your fat cells is an evolution-based survival mechanism; store it now for later use when food is scarce.

Would it be gauche of me to say I find that proposition absurd?

For that assertion to be factual - for that to be evolution-based - our ancient ancestors would have had to have been living in glucosis rather than ketosis for millions of years. That was not the case.

As discussed in chapter 5, there is no rational basis to believe that prior to several thousand years ago mankind lived in glucosis, and there is every reason to believe he existed for millions of years in ketosis.

Because man existed in ketosis his diet was almost entirely protein and fat.

As mentioned a moment ago, fat intake does not elevate insulin from baseline.

Protein consumption in significant quantity does produce a modest insulin response.

NOTE1: *Because protein does not significantly elevate blood glucose, researchers are unable at this time to explain how protein initiates insulin secretion.*

Understanding that protein elevates insulin moderately from baseline, let us consider how these elements play out when in ketosis.

At this point I need to share with you something some modern 21st century humans often find uncomfortable. Our ancestors ate the fattiest part of their kill first. The part we value today - the lean muscle - was considerably less important to ancient man. If the animal being eaten was sufficient of size (or quantity) that the man was satiated by the fattiest parts of the animal, the muscle portion of the carcass was left for scavengers.

The reason man preferred the fattiest parts of the animal is that the fatter the substance the more energy it provides. A gram of fat provides twice the energy of a gram of protein or carbohydrate. All meat-eating animals, including our ancestors, consume the fattiest parts of the animal preferentially.

Let us consider the physiological effects of a meal consumed by our ancient ancestors in ketosis. For this example let us say the macro ratio of his animal flesh meal was 50% fat, 50% protein. (There would be no carbs.)

The 50% protein would produce a modest elevation of insulin, signaling the cells to open up and receive nutrients circulating in the blood.

Fat consumption would not induce any elevation of insulin.

The nutrients flowing through the blood would be triglycerides from chylomicrons - which are not dependent on the modest insulin increase for uptake - and the amino acids [what protein is broken down to during digestion] - the cellular uptake of which is dependent on elevated insulin.

Do you notice what is completely missing in this physiological equation? If you said, "Any demand for the liver to convert excess glucose to triglycerides and store them"; i.e. any activation of the HLS, you nailed it!

Since our ancient ancestors lived in ketosis for (at a minimum) millions of years - without any need to convert and store that which was not in their diet to any extent (carbs) - how would this mechanism have established itself for the purpose ascribed to it in the modern narrative? Answer: It would not have. And it did not.

There is a far more rational explanation for the existence of the HLS mechanism.

Given the diet of our ancient ancestors, there would be virtually no occasion in which blood glucose would rise significantly above baseline; it routinely stayed low, stable, and healthy.

Imagine one of your ancient ancestors in ketosis. It is Springtime and he stumbled across a plant with several ripened high-carb edibles, such as an ancient version of artichokes, which are extremely high in carbohydrates. He eats several of them. Not long thereafter his blood glucose level would become astronomically high.

> **NOTE2:** *As anyone who has eaten a high-carb meal while in ketosis can attest, the sudden influx of carbs produces a very unpleasant physical reaction; anywhere from the shakes, bloating, and lethargy, to vomiting. Point being, ancient man would have learned what not to eat to avoid that experience.*

As discussed earlier, prolonged high blood sugar is toxic for the body's internal organs and tissue, so his body swings into action to reduce the blood glucose level.

This is where things get interesting.

Having lived his life in ketosis, your ancestor's cellular mitochondria had no use for glucose. That is a phenomenon today referred to as "glucose rejection". 60,000 years ago it would have simply been...normal.

Despite the cells having little preference, desire, or need, for glucose, if the glucose level became significantly elevated something had to be done to bring it back down to baseline.

As discussed previously, the dramatically elevated insulin resulting from high blood glucose signals the liver to initiate conversion of excess blood glucose into triglycerides, as well as to store the newly manufactured triglycerides in adipose tissue.

Because his liver converted the potentially harmful excess blood glucose into fatty acids - the very substance almost every one of his 100 trillions cells use as their primary energy source in ketosis - most of the triglycerides, if not all, would have been oxidized promptly. It is unlikely much, if any, would have been stored.

That said, any stored fatty acids would be released from the adipose cells and transported to other (non-adipose) cells over the next 12 to 24 hours [between meals and/or during sleep] where they would be oxidized for energy.

At this point, let me ask you a question...

Does the synthesis of excess glucose into triglycerides, and its storage in the adipose tissue, seem like an emergency response to a rare but potentially harmful human mistake of consuming too many carbohydrates... or...an evolutionarily established mechanism to 'routinely' address a situation that would have been exceedingly rare for our ancient ancestors?

Let me phrase this to you the way it plays in my mind - the way I would say it to a friend in casual conversation; the HLS isn't a mechanism the body ever intended, or intends today, as a routine high blood glucose management system. It is an "OH SHIT!" mechanism available to resolve a dangerous situation, on an emergency basis, that would have been rare in the lives our ancient ancestors.

In other words, the HLS does not exist as a physiological tool to manage routine high levels of blood glucose and store it away for later, as you've been led to believe.

The HLS exists as an emergency system the body can employ on the rare occasions man consumes an unhealthy amount of carbohydrates.

The problem mankind has run into is a 12,000-year timeline in which humans have arced further and further away from the symbiotic Earth/Man relationship concerning what nourished man for, at minimum, millions of years.

> *"Keep your blood clean, your body lean, and your mind sharp."*
> *~ Henry Rollins*

This 12,000-year arc moving away from the symbiotic Earth/Man nutrition relationship is the result of 'technology'. Or more pointedly, the way man has chosen to utilize certain technology.

Let me be clear; technology is a beautiful thing. It has significantly improved the quality of life for billions of people in innumerable ways. That said, I'm sure you are aware technology can be used for benefit or harm. We can, and do at times, use it in ways that are adverse to our own best interests.

The technology timeline that has had mankind arcing away from the symbiotic Earth/Man nutrition relationship looks like this:

First Agricultural Revolution - Began 12,000 years ago. Marks the beginning of the technology-base transition of mankind from hunter-gatherer to having increased autonomy over what he would eat via farming and animal domestication.

Second Agricultural Revolution (circa 1650 to 1890) - Comprised of a series of small "revolutions" in agricultural technology. Dramatically increased the output of farms. During this short period wheat production rose from 10.45 bushels per acre to 26.69 bushels per acre.

Manufactured Food Revolution (1900 to today) - Increase in farm yields were coupled with emerging science to dramatically alter the nature of man's food supply. For the first time in history man's food was being altered by researchers in laboratories and manufactured in factories.

NOTE3: *"Manufactured Food Revolution" is my title for this technological development. I was unable to find any term assigned to it. Another odd omission.*

The result of these changes - particularly the Manufactured Food Revolution - has been to allow man to increase his intake of foods containing carbohydrates to virtually any level. Within a couple of thousand years mankind went from a carbohydrate intake in the realm of 0 to 2% or 3% of total daily caloric intake, to essentially any amount one chooses.

In the 21st century, the average American consumes roughly 50% of his/her calories from carbohydrates. Some, much more.

For at least **tens of millions of years** humans consumed a minuscule quantity of carbohydrates. That means man's genetic coding - how the body is hardwired to operate - is based on **that** construct.

Then, suddenly, in what is in comparison to tens of millions of years, **the blink of an eye** man's daily consumption of carbohydrates leaped to 50% or more!

> *"The major problems in the world are the result of the difference between how nature works and the way people think."*
> ~ Gregory Bateson

If we agree the HLS is, in its nature, an emergency response to a rare occasion of high blood glucose, and not a daily high blood glucose management system, what might be the ramifications - the consequences - of increasing daily carb consumption from that of our ancient ancestors (0 to 2% or 3%) to a whopping 50% or more, creating what is, from the perspective of our genetic code, a daily - and relatively constant - blood sugar '**emergency**'?

Would forcing the body to employ an emergency mechanism - intended to be used perhaps a few times a year - to constantly act as your body's "high glucose management system" produce consequences?

If so, what might those consequences be?

Would the consequences be immediate? Or might they occur gradually from a deterioration of the body's ability to effectively manage the problem on a daily basis - over decades?

Would the consequences be immediately noticeable? Or might they grow quietly, beyond our awareness, while all appears outwardly normal?

Did I mention that some researchers believe 75% of the American public is currently in some stage of insulin-related health problems, even if not discernible to them at this time except by very specific testing that is rarely done?

CHAPTER EIGHT

TURNING A BLIND EYE

"I'm just gonna hope that science advances faster than I can deteriorate.
~ Ronda Rousey
(First female inducted into the UFC Hall of Fame.)

In the chapter quote (above) we readily understand why Ms Rousey would be concerned about science advancing fast enough to off-set the abuse her body endures. It is obvious to us because we can see the 'cause and effect' right before our eyes.

There is a reason professional fighters have a limited number of years in the game. They are expecting their body to perform, year-in & year-out, a function (physical combat) the body was never intended to engage in routinely over time.

Everyone understands and acknowledges that reality. It makes sense to us.

Yet when other forms of physical deterioration take place, subtly, gradually, over a lengthy period of time, in a way that is concealed from our sight, we are remarkably resistant to acknowledging it is occurring.

Furthermore, if acknowledging deterioration is taking place might suggest changing our well-established preferences and behaviors, the vast majority of people will choose not to see it, even if/when clearly pointed out.

After all, if we simply reject reality its consequences will go away, right?

You might think me a cynic. When it comes to human beings effecting change in their lives, when logic and reason dictates doing so, I will admit observations of my fellow man haven't left me beaming with optimism.

I have a wonderful friend who is a cardiologist. In the early days of his practice he attempted to educate patients. He no longer does so unless something a patient says leads him to believe the patient values facts.

He used to explain to all his patients that heart disease doesn't just happen; it is (in the vast majority of cases) created through choices/actions. His remarks focussed on helping them understand they could prevent, reverse, or mitigate the situation by changing their habits.

One of the more disturbing things he shared with me is that because virtually no other medical doctors, whether general practitioners, cardiologists, endocrinologists, etc. have ever told the patient that getting heart disease is all about the personal choices we make, his patients have refused to believe him; responding with, "None of my other doctors say that."

> *"At least half our patients cannot be persuaded that they must permanently alter their eating habits to save their lives."*
> ~ Dr Robert Kemp (1963)

He estimates only about 1% to 2% of patients have any interest in hearing the truth. Most want something else. Anything else. And they're not particular about what it is, just as long as they don't have to change anything in their lives.

When I was studying law I came to understand something that was, for me at that time, shocking. I came to understand the job of corporate counsel is not to ensure the corporation for which they work complies with the law, as had been my naive presumption. I came to understand the true value of corporate counsel is in advising company executives how to violate the law whenever it suits the interests of the business without the company or its officers being held legally accountable.

Likewise, patients aren't interested in changing their habits in order to stay healthy, or return to a healthy condition if ill. What they desire from their physician is a means by which they can continue making unhealthy choices without having to pay the price. They want to continue doing what they have always done, just absent consequences.

In the minds of these patients doctors are not actually "medical practitioners", as we understand that phrase. The unspoken role into which these patients thrust their doctor is little more than a "consequence eliminator".

It should be manifestly obvious that eliminating or forestalling 'consequences' is not remotely the same thing as being healthy. The evidence strongly suggests that "being healthy" is not the agenda of most patients.

As my cardiologist friend shared with me, "If heart disease patients cared about the facts concerning what is causing their disease, and acted to correct the factors underlying its causation, the sale of statins would drop to zero. But patients just want the statins."

Acknowledging that isn't cynicism. It is realism.

What might be the possible 'consequences' I mentioned at the end of the last chapter?

Allow me to answer in the form asking another question.

How is it that in our modern scientific world, in which we have spent tens of billions of dollars - perhaps even hundreds of billions - on research concerning heart disease, cancer, diabetes (types 2 & 1), insulin resistance, hypertension, hyperinsulinemia, Syndrome X, Alzheimer's disease, and on and on, all of these diseases are proliferating at an ever-increasing rate?

How is it that as the scientific community learns evermore, and we have ever-better technological tools, and vastly more of them, with which to 'fight the good fight', we are **losing** the fight - **big time**?

As I am writing this the medical industry occupies 18% of the U.S. economy. Yet the health of Americans continues going downhill.

Almost 20% of the total U.S. economy; yet an ever-greater percentage of Americans are getting sick. People are getting chronic diseases younger in life than ever before, and that trend is accelerating.

In light of these facts, is it not rational - even imperative - to question whether we're missing something - something critical?

It seems to me that to not ask that question repudiates logic and reason.

Research shows most Americans have been following the nutritional advice provided to them by the U.S. government and the medical industry over the last 50 years. (The medical industry adopts uncritically whatever the U.S. government puts out and regurgitates it to patients.)

So...not only do we have the best medical technology mankind has ever known, and a significantly greater volume of it than at any time in history, plus a level of money spent on research that boggles the mind, but we have a fairly obedient public doing what they are told. Yet the health of Americans has gotten worse every year - for at least 40 years running.

If we ponder where the problem may lie, would it not be reasonable to look for some issue the government and the medical industry have (for whatever reason) never considered - or at least been unwilling to discuss publicly? (I do not at this time raise the question of 'why' the govt and medical industry may have avoided considering certain matters.)

Let us recall this definition from chapter 2; most specifically "NOTE1".
Glucosis - A state in which the 100 trillion cells of the body use predominantly glucose as their source of energy.

> **NOTE:** *There are only two "hemispheres" concerning which type of 'fuel' cells use for energy, One is glucosis, the other ketosis.*

> **NOTE1:** *Glucosis is not a recognized medical term. Oddly, the medical community has never fashioned or attributed a name for the mode in which the body uses glucose for energy. I created the term "glucosis" to resolve that oversight.*

Does it not seem more than a little bizarre that the medical industry, or government agencies such as the National Institutes of Health, have never put a name to a very specific mode of fueling the cells of the body;

the mode which the bodies of probably 99.5% of Americans use?

This failure would not be quite as bizarre (though still pretty bizarre) had they not given a name to the other mode of fueling the cells. There are only two modes. They named one and left the other unnamed. For 100 years.

NOTE1: *I say "For 100 years" because the phrase "ketogenic diet" was coined by Dr. Russell Morse Wilderin (an early expert in carbohydrate metabolism and metabolic disorders) in 1921. The state of ketosis was identified shortly before Dr. Wilderin coined the phrase identifying the diet that would initiate ketosis.*

For 100 years no one in the medical community thought to give it a name? A 100-year oversight; a 100-year "oopsy"?

Does it not seem almost beyond comprehension that I am the first person naming it - finally - in 2019?

Perhaps we could understand this 100-year 'oversight' if it was merely the public - average Americans - who failed to give it a title. But that sort of thing is not their job; not their field of expertise. The failure is not theirs.

The failure rests squarely with the medical research community, who, in every other matter, have a well-established propensity - one might even say, a passion - for naming even the tiniest things.

You may recall in chapter 2 I said, "One of the things that slowed my initial progress in learning physiology was that **everything** has unfamiliar medical or science-specific names." I emphasize "everything" here to make the point that they do, in fact, name **everything.**

Everything…except this one thing.

Naming every little thing is not merely a reality in the medical research field. It is a reality in all areas of science.

As just a few examples….

Astronomers name tiny dots in space they believe, but cannot yet prove, are planets or distant solar systems.

Physicists have named sub-atomic particles, though such objects are not remotely on the mental radar of 99.99% of the public.

Scientists name things they have never seen, only theorized, such as quarks.

In science and research everything must have a name because if not named how can a person communicate about it with his/her peers?

> *"The world hasn't been fully seen, until it is named."*
> *~ Lynne Tillman*

Should I raise the issue that it is virtually impossible for the public to be aware of - no less debate the merits/demerits - that which has no name?

I'm sure I'm just being silly. The failure to give it a name couldn't be anything like that, right?

Whatever reason one may attribute for this 100-year failure, I have rectified it with the name, "glucosis".

NOTE2: *Do not confuse "glucosis" with "glycolysis". Glycolysis is the pathway by which glucose is converted into energy within a cell. It does not identify the mode of overall body reliance on glucose as its primary energy source, which is what "glucosis" defines.*

With all that said, are you aware there has never been a shred of research done comparing the overall health of people whose bodies are operating in glucosis with the health of people whose bodies are operating in ketosis?

NOTE3: *There has been small independent "evaluations" (involving questionable objectivity) of the "possible" benefits of being in ketosis for very short periods of time; measured in weeks. There has never been a large-scale well-funded head-to-head study of these two opposing systems.*

NOTE4: *NIH has done some limited studies only for the purpose of determining if the "Keto diet" (which is not synonymous with the critically important phrase, "living in ketosis") is safe for weight loss. This is of course not the head-to-head comparison between glucosis and ketosis that is so desperately needed.*

Considering that scientists, nutritional researchers, and medical doctors are aware of every issue I've raised with you in the previous pages, how do we explain this…um…oversight?

Again, there are only two modes of fueling the body's cells; glucosis and ketosis.

Yet…somehow…as America's health has continually deteriorated over the last 60 years, it never occurred to anyone in any medical or nutritional research organization, or the U.S. government's health research apparatus, that perhaps these two systems should be subjected to a head-to-head competition (so to speak) to see which produces the healthiest participants?

Given the billions upon billions of dollars spent on nutritional research over the last half century, does anyone honestly believe the failure of the medical and nutritional research industries, and/or various government agencies, to perform such a simple and obvious comparison is merely an oversight?

CHAPTER NINE

NO SCIENCE NEEDED (FOR LIES)

"Can a geology teacher blithely tell his students that the earth is flat…? That's not academic freedom, but dereliction of duty…"
~Jerry Coyne

My fellow journalists who've written on the subject of how the nutritional research industry has operated over the last 60 years are…generous…in their acceptance of reasons for the endless lies, distortions, and misrepresentations that have come flowing out of that community.

They appear to accept the excuses proffered (almost always by retired researchers and industry executives) to explain why the lies were told, and continued to be told even when it was clear the assertions were not credible; excuses such as (paraphrased)…

"We were attempting to act in the best interest of the public", or…

"Saying something is better than not saying anything when people are getting sick and dying", or…

"Detailed specialization prevents researchers from seeing 'the big picture'", or…

"The cost of thorough well-performed research is simply too high, so we ran with what we had even though it hadn't been proven factual."

I am not so generous.

I have spent much of the last 20 years researching pervasive societal myths that are a detriment to society. They always exist by intent.

Propaganda on a subject was so successful it eventually turned a falsehood into a widely accepted societal myth Americans now believe to be true. That propaganda has always benefited certain groups/entities.

This is no less so with the seemingly endless list of falsehoods put out by the nutritional research industry and the U.S. government concerning diet and nutrition.

Make no mistake; many of the bogus hypotheses put out by these charlatans are still believed to be factual by a huge percentage of the American people, even decades after being disproven.

The original false assertions received massive media coverage. The outcome of clinical trials disproving those assertions received almost none. But I don't want the guilt to rest with the media. While the media has certainly been co-conspirators over the decades, the primary guilt rest with those who have control over "studies".

I have compiled a list of more than 20 different ways researchers can "cook" the results of a study. These range from designing a study from the ground up to render a pre-determined result, to burying 'inconvenient' findings [i.e. facts that would embarrass earlier researchers and benefit the public] hundreds of pages deep in a study report, dramatically reducing the odds anyone will find/see it.

Another nifty trick, when research results contradict the narrative they desire to promote, is to either not publish the findings [Hey wait! Isn't this supposed to be science?!] or publishing them in some inauspicious journal no one with enough clout to draw attention to the 'inconvenient fact' would read.

There are many more such corrupt practices employed to ensure study results - scientific facts - don't get in the way of the desired industry narrative.

> *"A colleague once defined an academic discipline as a group of scholars who had agreed not to ask certain embarrassing question about key assumptions."*
> ~ Mark Nathan Cohen, Health and the Rise of Civilization (1989)

The foundations of, as well as past and/or on-going promotion of, these destructive false narratives are always rooted in money, personal advancement, the establishment or maintenance of reputation, and/or ego. And make no mistake; those things are what is important to those in the nutritional research field, not your health.

Time and again I have observed how people who try to alert the public that they are being lied to by those who owe the public a duty to tell the truth, are attacked, vilified, their careers often destroyed, while those who lend their prestige to the lie - who speak glowingly in support of it - are rewarded with professional advancement, stellar reputations, and significant increases in wealth.

This is not to say that everyone who may have gotten something wrong is a dirty low-down so-and-so. But when you read the history of the nutritional research industry over the past 70 years it is child's play to track the ascension of those who unabashedly mischaracterized hypotheses as 'scientifically proven fact' to the public, even after the overwhelming weight of evidence made clear the hypothesis was, at best, highly dubious, and at worst, flat-out wrong.

Let me go a step further. It is not only that these false narratives pushed the American people into becoming sick and dying early, but just as when the criminal justice system convicts an innocent man of murder, the real killer is still out there! And so it is here. When lies are proffered, and accepted, as proven facts, from where then comes the impetus - and the funding - to find the real killer?

THE KEYS EXAMPLE

Ancel Keys (1904 - 2004) was brilliant...and thoroughly corrupt.

Keys attained a level of notoriety, respect, adulation, and wealth previously unheard of in the nutrition research field. His corruption, which he kept hidden from the public by being a masterful political hand within the upper levels of the nutrition research and advisory community, set an example - established a model - for others in the field to follow over several decades.

What those in the nutritional research industry saw when looking at Keys was a man with (by all accounts) a compelling & forceful personality, great intellect, who built for himself what no other nutrition researcher had ever created.

- No other nutritional researcher had ever been a household name.
- No other nutritional researcher had graced the cover of Time Magazine.
- No other nutritional researcher frequently appeared on TV.
- No other nutritional researcher had attained celebrity status.
- No other nutritional researcher had created such wealth for himself.
- In the nutritional research world, Keys was as close to a god as had ever been.

And Keys built all of that by lying to the American people.

In the late 1950s and early 1960s Keys created what would come to be known as the "diet-heart hypothesis". Because I'm not keen on vagueness or ambiguity, I call it by a more descriptive phrase; Keys' "dietary fat will kill you hypothesis".

Keys (and his sycophant industry allies) used his force of personality, his intellect, and his political skills, to get his hypothesis accepted as "valid" by the American Heart Association, and not long after, by the U.S. government.

In 1980 the U.S. government turned Keys' hypothesis into a perceived "scientific fact" by issuing dietary guidelines that called for reducing dietary fat intake, with special emphasis on saturated fat (hereafter, "sat-fat") in order to (allegedly) prevent heart disease.

Keys was also the most significant and influential figure in having dietary cholesterol declared a major public health risk.

Keys and a small cadre of powerful and well-placed co-conspirators dominated the discussion about the cause of heart disease for three decades.

He and his allies were so powerful that while an entire nation tried its best to change their diet in order to follow the dietary "facts" put out by the U.S. government (and parroted uncritically by the medical industry) no large-scale public clinical trials were conducted to test whether Keys' "dietary fat will kill you hypothesis" could, or would, withstand scientific scrutiny.

Such was the power and influence of the "Keys Cabal" that subjecting his hypothesis to the rigors of scientific testing was considered unnecessary - even insulting and ungentlemanly.

Those who pushed for meaningful testing of Keys' hypothesis saw their careers ruined. Those doctors or researchers who had the audacity to question Keys' hypothesis within the nutrition research community were mocked and ridiculed by the powerful Cabal members who occupied top position in the industry and quickly retreated - at least if they wanted to work in the field in any meaningful capacity in the future.

This is not to say that no research was done at all. A number of studies were done over the years looking at various incremental particulars that would have to be found factual if Keys' hypothesis was valid. Every one failed to provide the evidence necessary to support Keys' hypothesis. In other words, as Keys and his Cabal continued to influence the dietary habits of Americans from New York to San Francisco, the underlying fundamental science required to substantiate Keys' hypothesis could not be established.

You can probably predict how this story ends. In the first few years of the 21st century several clinical studies came to fruition which thoroughly debunked every single one of Keys' assertions.

Furthermore, this wasn't just a case of a potentially valid hypothesis being given too much credibility. Keys cooked his data from the start.

We learned from one of Keys' early research assistants (years too late) that when Keys was compiling his original data in Europe, using survey forms, if he found responses from the participants in a particular locale were incompatible with his hypothesis, he simply threw the survey forms in the trash, telling his assistants the respondents had obviously lied.

Keys wasn't looking to "test" his hypothesis, as proper science would dictate. He was only interested in convincing the world it was factual - that he was "right" - no matter what the evidence dictated.

But it's worse than that.

Keys did research in 22 countries. When he published his second report, entitled "Seven Countries Study", he incorporated the data from only seven countries. Unsurprisingly, they were the seven countries the data from which supported Keys' hypothesis. The data from the other fifteen countries was excluded because it would have undermined his agenda, which was to be "right", no matter the facts.

Facts that would have destroyed Keys' hypothesis, he simply hid from the American people.

There is much more information available concerning Keys' lack of integrity, but I'm sure you get the point.

It is important to take particular note of this fact: Keys' hypothesis was wrong all along. And it would have been easy enough to prove it so had there been the will to hold it up to scientific scrutiny. No one with the will had the power. Those with the power were Keys' confederates.

Keys was also a friend of the powerful sugar industry.

In the early 1950s the sugar industry began doling out "research" money to prestigious universities across America. Its first bribe…I mean "contribution"…(adjusted to 2020 dollars) was $30,000,000, allocated between several schools.

Never being one to pass up becoming the recipient of industry largesse, or miss an opportunity to take a stance against facts, science, the health of the American people, Keys was quick to become an outspoken friend of the sugar industry. For the next three decades he was a powerful and respected voice in the national debate over sugar, telling the American people sugar was perfectly safe and not a health problem.

Note I said the sugar industry started dispersing these funds (i.e. bribes paid for corrupt research results) in the early 50s. Do you think it coincidence that Keys began his multi-nation "research" in the early-50s? It is not.

Keys' multi-nation "study" was funded by the sugar industry!

In other words, at a time when the sugar industry perceived itself under attack from the emerging concerns/questions about potential health risks associated with sugar consumption, Keys took the industry's money and dutifully produced a report concluding America's heart disease problem was not the result of the product his benefactors sold, but rather the result of a substance mankind had been consuming without adverse health issue for millions of years.

> *Science advances one funeral at a time.*
> *~ Otto Heinrich Warburg*

From the early 60s through...well...this very day, Ancel Keys has likely been responsible, single-handedly, for more needlessly premature deaths than all the wars of the 20th Century, combined. Quite an achievement.

He was also a role-model.

Several generations of researchers learned by his example that rewards in the nutritional research field come not from ascertaining and publishing scientific facts, but from being able to put together a network of allies who can promote and support whatever you come up with that gains traction, facts notwithstanding. They learned that if you can control the narrative, you can have success. No science required.

Ancel Keys' fraud-based success set the moral tone for the next 40 years of nutritional research in the U.S., while hundreds of millions of Americans got sick needlessly and died prematurely.

THEY'RE CORRUPT. WE WIN ANYWAY!

Occasionally when I'm attempting to illustrate a point in a writing or video I will offer an experience of my own as an anecdote. I always preface the personal story with words such as, "I know my story is merely anecdotal...". In other words, I let the reader/viewer know I am aware that a personal anecdote does not prove a premise factual or correct. An anecdote is only evidence that I had a particular experience with a particular outcome. It is not evidence that others will, with the same experience, have the same outcome.

Yet the contextual meaning of "anecdote" is the individual, not a collective.

What do we call it when 1,000,000 people engage in a particular experience and all have the same outcome? While singularly the story of each person is an anecdote, who is it that believes we can, or should, dismiss a consistent mass-outcome of millions of people as "merely anecdotal"?

The point being; 1,000,000 matching anecdotes (absent a significant number of non-matching results) cannot be dismissed if we are people of logic and rational thought.

Collectively, those 1,000,000 matching anecdotes become something else.

"The plural of anecdotes is evidence."
~ George J. Stigler, Nobel Prize in economics

In the research industry, when researchers pay close attention to a population base (without intervention or control), with an eye to specific issues such as environment, lifestyle, nutritional characteristics, etc., for the purpose of determining result trends, that is called an "observational study".

Observational studies are among the lowest cost forms of research because they rely on interaction between medical personnel and the population that is already occurring. (Those whose particulars are being evaluated aren't even aware a study is taking place.)

In other words, an observational study simply relies on data acquired through routine interaction. Nothing special required.

One might wonder why, in a nation as populous as the United States, no research organization, public or private, has engaged in this type of simple, low-cost study, comparing the health of those living in glucosis with the health of those living in ketosis.

> *"For a scientist, this is a good way to live and die...*
> *excitedly finding we were wrong and excitedly waiting*
> *for tomorrow to come so we can start over."*
> ~ Norman Maclean, Emeritus Professor of Genetics,
> University of Southampton

That's an excellent question, which I'll answer in the next chapter. But at this time I will simply share the reality that no such studies exists. And I'm not holding my breath for the nutritional research industry to conduct any such studies. Ever.

Why?

Because they already know what the outcome will be - and too much is on the line. (We'll get to that shortly.)

What happens when the research industry is controlled, corrupt, and refuses to engage in even low-cost observational studies that would provide much needed clarity on a critical issue; an issue that could potentially save hundreds of millions of Americans from disease and early death?

Answer: We move on without them. They've rendered themselves irrelevant.

If there is an overarching purpose behind this book it is to inform you that the nutritional research industry has failed us on this subject - completely.

More importantly...that failure is **intentional.**

The good news is we already have the needed evidence.

Hundreds of thousands, perhaps millions, of people in America are living in ketosis right now. Tens of millions worldwide. And the results they are experiencing are universally positive and overwhelmingly similar.

That **is** evidence!

CHAPTER TEN
THE TRUTH AIN'T PRETTY

A foolish faith in authority is the worst enemy of truth.
~ Albert Einstein

America's nutritional research institutions, public and private, are bought and paid for by special interest groups, most notably (but not solely) the billion-dollar corporations comprising the processed food industry - and of course Big Pharma. This has been the case for many decades.

> "The medical profession is being bought by the pharmaceutical industry, not only in terms of the practice of medicine, but also in terms of teaching and research."
> ~ Arnold Seymour Relman (1923-2014) Harvard professor of medicine and former Editor-In-Chief of the New England Journal of Medicine

In September 2016 the New York Times ran a piece by investigative reporter Anahad O'Connor detailing how, back in the 1960s, the sugar industry bribed researchers at Harvard University to write a report in such a way as to falsely and corruptly shift the blame for various health problems away from sugar consumption.

That corrupt influence continued for decades.

One of the Harvard conspirators, nutritionist D. Mark Hegsted, went on to become the head of nutrition at the United States Department of Agriculture, where in 1977 he helped draft the forerunner to the federal government's dietary guidelines. The USDA also had in its nutrition department another of the conspirators, Dr. Fredrick J. Stare; Chairman of Harvard's nutrition department.

This decades old story of successfully bribing researchers to provide false conclusions is only the tip of a very large, very nasty, iceberg. Just one of the countless Ancel Keys-type stories of corruption from America's nutritional research industry.

> *"They say 'you really need a high level of proof to change the [nutritional] recommendations,' which is ironic because they never had a high level of proof to set them."*
> ~ Dr Walter Willett, Harvard Professor

The most famous iceberg may be the one that sank the Titanic, but the iceberg of U.S. nutrition research fraud - of which so few are aware - has been consistently (and quietly) sinking the health of our nation by destroying the health of its citizens on a scale never before seen in human history.

Private research organizations rely primarily on corporate funding. Sure, these days they are required to disclose their funding sources, and researchers must file conflict of interest statements, but virtually no one - most especially the American public - reads the statements or compares them to the research performed and results reported.

Nary a week goes by I don't see at least one or more articles with headlines telling us about the latest and greatest nutritional research "revelation". When I read the articles (which usually omit a link to any study), far more often than not the claims are utter nonsense. Even the most rudimentary knowledge of physiology debunks them in a hot second. Yet the American people, the vast majority of whom have no knowledge of physiology, are endlessly bombarded with stories about one nonsense study after another.

Ironically, even the giant processed food conglomerates have to deal with this tsunami of false research claims. Those enterprises, who need solid science from which to develop their latest and greatest products have found that in 43% of cases their in-house laboratories - which are the best money can buy - can not replicate the results detailed in published scientific study reports.

Think about that. We're heading toward half of all claimed research results being either fraudulent or the result of ineptitude!

But you don't have a multi-million lab staffed with trained research specialists to tell you what results are factual and which are false. So how do you know?

Surely the U.S. government can't be corrupted by these gargantuan corporations, right?

OK...you caught me; I couldn't even write that with a straight face.

Corruption happens because there is impunity.
~ Joao lourenco

I'm confident most Americans are aware Big Business exercises a disproportionate level of influence in Congress. In fact, over the last several years three major U.S. university studies concluded that in more cases than not, when a majority of Americans disapproved of a proposed law, but billion dollar industries support it, Congress enacts it.

While Big Business influence over Congress is on full public display, how much more control do you think billion dollar industries exert at the agency level, where what happens is not, in any meaningful sense, open to public scrutiny?

There is a small group of high-ranking government officials in the National Institutes of Health (NIH) who decide what health-related research is funded - or not. They are, in the main, cut from the same cloth as Ancel Keys. Furthermore, while they are, technically, government employees, they have far more in common with a President or Vice-President of a nationwide medical insurance company than they do traditional government employees.

Research informs us that those who are duplicitous and manipulative are more likely to attain top positions in large bureaucratic organizations than those who are forthright and honest. Couple that with the culture of corruption existing in the nutritional/health research industry for the last 60 years and you may begin to understand the depth of the problem.

Are there nutrition/health researchers who want to do good; want to provide the scientific facts and let the chips fall where they may? Of course. But those at the top make sure those researchers are never allowed to touch anything "dangerous". By "dangerous", I mean disadvantageous to the economic interests of various industries/entities/persons.

Let me be straight with you....

If factual scientific results from a proposed study could reasonably be anticipated to cause economic harm to various industries/entities/persons, those trials will not be funded. Whether such trials might save hundreds of millions of lives isn't even a part of the equation for those in control of nutritional/health research industry funding.

In chapter 9 I discussed the benefit - the wisdom - of conducting a study that would compare the totality of health of Americans living in glucosis with the totality of health of Americans living in ketosis. I told you no such research has been done, nor do I have any expectation such research will be done. Ever.

No such research will be permitted - will be funded - unless or until there is a house-cleaning of all senior executives at NIH. The current executives are there to protect the revenue of various industries, not your health. I see no scenario in our future that would result in that house-cleaning occurring.

In the last 60 years or so America has moved from a manufacturing economy to what economists call a "knowledge economy." In other words, unless you know something that is of value to business you will not be upwardly mobile.

Getting out of high school and "working your way up" is no longer a thing. When you apply for a job today you must already possess 'knowledge' from which a company perceives it can benefit.

Since the nature of the economy is directly tied to our survival - or at least our material happiness - society tends to conform itself to the nature of the economy. If we have a 'knowledge' economy, in turn we will develop a 'knowledge' society.

In the U.S. version of a 'knowledge economy' - or 'knowledge society' - what constitutes 'knowledge' is determined by "the authorities".

As an example, a large company is unlikely to hire you if you tell them the knowledge you possess relevant to the job for which you are applying was developed from extensive self-study.

By contrast, if you applied to the same company, for the same job, with a degree from an accredited big-name university, your odds of getting hired would be exponentially greater.

But here's the thing; the person who engaged in extensive self-study may possess more useful/applicable knowledge, clearly has shown him/her self to be highly self-motivated, and may turn out to be a stronger performer than the formally educated applicant. But we'll never know. The self-educated person won't even be considered.

The point of this example is that society's definition of 'knowledge' today is limited to information provided by "the authorities". Only a seal of approval from '"the authorities" makes information real, true, or…**valuable.**

Whether 'knowledge' is factual is irrelevant. Only when it is provided or approved by "the authorities" is it considered to have **value.**

NOTE1: *The exception being the tech industry, wherein self-education and dynamic innovation is the culture.*

Wikipedia has become the go-to knowledge base for millions. Yet, sadly, Wikipedia is the poster-child for supporting anything and everything "the authorities" say - no matter how plainly non-factual - while disparaging any information that runs counter to what the authorities say - no matter how obviously factual.

In short, while Wikipedia may be useful for mundane inquiries, if one is seeking factual data concerning an emerging truth that is opposed by an establishment lie, Wikipedia is worthless.

With that in mind, let us return to the matter of research comparing the health of people living in glucosis with the health of those living in ketosis.

There isn't any - at least by "the authorities".

If we live in a society in which 'knowledge' is considered to have value only when provided or approved by "the authorities", what then is the ramification when the authorities remain silent on a subject? Answer: There is nothing considered 'knowledge of value' on that subject in society.

Think about the implications of that. How "1984"!

How much factual knowledge do people trained and 'approved' by "the authorities' possess? That is obviously dependent on the person and his/her commitment to facts, but let me share an anecdote that illustrates how dire is the situation.

The research I did for this book has been significant. In many instances I went down the 'rabbit hole' I mentioned in chapter 3. At times I went to a depth where the information I sought is unknown at this time, or finding it was like the proverbial needle in a haystack.

In one matter I felt I could not be absolutely certain about a sequence of physiological events unless I tracked down a particularly arcane point of chemistry regarding lipid transport in a very specific situation. I thought it would be simple enough to find the information. I was wrong.

Hours of internet searching turned up nothing. I quarried several medical doctors. They didn't know. I asked others holding various licenses in the medical field. I even reached out to a person some consider the most knowledgeable lipid expert in America. Nothing.

During all of those inquiries there was someone to whom I wanted to direct the question. Yet I put him last on my list because I know he is incredibly busy, and I don't know him personally - though I am well familiar with his work.

His name is Dave Feldman.

Dave is a "citizen scientist". He holds no medical licenses and has never been to medical school. Yet Dave has more practical (as opposed to laboratory/theoretical) knowledge concerning lipids than probably anyone in the U.S., possibly the world

Of all those with whom I inquired, only Dave Feldman knew the answer.

But Dave isn't "the establishment". Dave doesn't earn his living from lipid research. He simply has a passion to know the facts rather than blindly accept the establishment's rhetoric. I can relate.

Not being part of "the system" allows Dave to do things that would never be permitted by the "Gate Keepers" of NIH or private research orgs.

Dave and I are both ardent advocates of "N=1"; shorthand for using yourself as the first and best research subject. (More on that shortly.)

Dave's N=1 research standards are vastly stricter than those used by the nutritional research industry for traditional research. His standards are, simply stated, exemplary.

Dave has used social media and his website [cholesterolcode.com] to create a worldwide community of involved participants/activists.

As an example, Dave will set out what blood test profiles he'd like to see - coupled to a particular dietary profile - and thousands of people from all over the world send him their data!

He is 100% transparent with what he is doing, and why. (A far cry from the U.S. nutritional research industry.)

From that ever-growing pool of participants, and the lipid patterns Dave has been able to observe therefrom, he is able to authoritatively discuss various lipid-related physiological realities - particularly about people living in ketosis - you won't hear from any establishment source.

Consider this...

If the nutritional research industry, and controlling forces such as NIH, were doing their job there would be nothing for people like Dave Feldman and myself to examine. The facts we pursue would already be known, and readily available in the public domain. They would already constitute 'knowledge' within the meaning of that word in this new (and odd) authority-dominated "knowledge society" in which we live.

Dave Feldman's work, and mine reflected in this book, fill a vacuum; a vacuum left when those who are reaping large salaries, prestigious titles, esteemed reputations, generous rewards from moving back and forth between industry and government, and cushy retirements, aren't doing their job.

But of course they **are** doing their job.

They are keeping the kind of information I address in this book from receiving even a penny of research funding. They are fulfilling their role as the Gate Keepers of the status quo, ensuring the numerous industries whose interests they serve are never financially placed at risk.

I find no constitutional authority for U.S. government to be involved in nutritional science or to make dietary recommendations to the American people. From 1791 until 1980 the federal government kept its nose out of that part of our business. The government's first official and public involvement began in 1980.

Would you be surprised if I told you the beginning of the dramatic increase in American obesity can be traced straight back to 1981?

> *"We'll know our disinformation program is complete when everything the America public believes is false."*
> ~ William Casey, Director of CIA (first staff meeting, 1981)

While I would greatly prefer government not be involved, the reality is not only is government involved, but it is the single most influential entity in America on the subject. And running true to form, it has misrepresented conjecture as proven science, outright lied about some of the most significant particulars, and acts as Gate Keeper to ensure no "official knowledge" is developed that would harm multi-billion dollar corporations.

What exactly is the science that must stay hidden lest its exposure harm multi-billion dollar corporations?

CHAPTER ELEVEN

FACTS.
THEY MATTER.

"The consequences of things are not always proportionate to the apparent magnitude of those events that have produced them…"
~ Charles Caleb Colton

What are the consequences to you, and to society, from the lies you've been told by the nutritional research industry?

That subject is so vast it could easily be an entire book unto itself.

If revealing the truth about nutrition and physiology would harm various industries, then we can assume people would alter their buying habits if the truth got out. Perhaps dramatically so.

Let's explore some physiology facts that would likely alter buying habits if known and understood by the American public.

We've all heard the phrase "chronic disease". But what does it actually mean?

According to the Centers For Disease Control (CDC) a chronic disease is a condition lasting 1 year or more that requires ongoing medical attention. Chronic diseases generally cannot be prevented by vaccines or cured by medication, they do not just disappear, and are not communicable.

It is estimated that 88% of Americans over 65 years of age have at least one chronic disease. At "55 and over" that figure goes to 78% having one chronic disease. Almost 50% have two chronic diseases. More startling is that 26% of Americans age 0-19 years have at least one chronic disease!

That said, there is no single controlling definition of chronic disease. Different organizations have shorter or longer lists of diseases they consider to be "chronic".

To keep this chapter relatively short, I'm going to limit the list of chronic diseases discussed to heart disease, cancer, type-2 diabetes, obesity, and Alzheimer's.

We'll also discuss precursors to a number of chronic diseases; insulin resistance and Syndrome X (metabolic syndrome).

NOTE1: *The researcher who first identified the specific combination of physiological abnormalities indicating significantly elevated health risk named it "Syndrome X." Years later, a person with more notoriety and clout in the research community branded it "metabolic syndrome". I believe the man who put the facts together deserves to name it, so to me it is "Syndrome X."*

Chronic disease has not always been a part of American life. In fact, before the 20th Century heart disease, cancer, type-2 diabetes, obesity, and Alzheimer's were exceedingly rare. Perhaps that fact should catch the attention of the astute observer.

Most people are familiar with at least the broad outline of what defines heart disease, cancer, type-2 diabetes, obesity, and Alzheimer's, so I won't take your time with that.

What is Syndrome X?

Although the physical abnormalities comprising Syndrome X have changed from time-to-time, and researcher-to-researcher over the last 100 years (well before it was formalized with a name), the current accepted definition is that a person has any 3 of the following 5 conditions…

1. Obesity of the adipose tissue
2. High blood pressure
3. High level of triglycerides in the blood

4. Low HDL
5. High blood glucose

Perhaps these 5 items make perfect sense to you in context of what we've discussed so far. If not, don't worry; we're going to give them a closer look.

After investigating how nutritional research has been conducted over the last 150 years, I observe that since WWII the 'real world' doesn't seem to exist for nutritional researchers. If something isn't seen by a researcher in his/her laboratory, it isn't "real" - even if everyone else on the planet can see it with perfect clarity.

As but one example, at minimum, hundreds of thousands of Americans are living in ketosis right now. If we combine numbers for the last several years, likely millions of Americans have lived in ketosis. Their universal experience is amazing health, being free of virtually all chronic disease, and its precursors.

Yet that real-world fact is meaningless to nutritional researchers. That reality doesn't exist for those researchers because it didn't happen in their lab. And it isn't happening in their laboratory because they can't get funding. No funding is available because once data exists from "the authorities", it becomes the 'knowledge' we discussed earlier; it has value.

I've also observed that in today's money/power driven research environment researchers have forgotten the doctrine of Occams' Razor, to wit; *when considering a theory or hypothesis, if the matter is susceptible to two or more explanations, the one that is least complex and relies on known natural factors should be presumed correct.*

That is exactly opposite of how today's nutritional researchers see it. After all, there is little money in that. The money, the ability to get "published", and the enhancement of reputation, is to be found in making one's hypothesis far more complex than required, even when a completely rational hypothesis exemplifying Ocamm's Razor is readily apparent.

Let's begin by taking a look at the precursors.

INSULIN RESISTANCE (IS)

I could literally write page after page about IS. It is an intriguing subject, in a nerdy physiologist kind of way. But let's keep it simple.

Insulin resistance is a condition in which the 100 trillion cells of the body exhibit a reduced sensitivity to insulin, and thus a reduced compliance with insulin's commands.

What does insulin command the cells to do?

If you recall from chapter 6, insulin commands the cells to "open up" and take in nutrients from the blood. Since insulin is predominantly a response to high blood glucose, the primary substances being taken into the cells in response to insulin's command is, of course, glucose.

When the cells develop and manifest reduced responsiveness to insulin's commands, it is called insulin resistance.

What does the body do to address this cellular disobedience - this 'rebellion' - to insulin's commands?

The body has innumerable feedback mechanisms. One such feedback mechanism informs the pancreas that the insulin level in the blood is insufficient to gain compliance from the cells. If the pancreas had a mind of its own it might think this odd because it just released the same amount of insulin it always has in the past, which successfully initiated the process of lowering blood glucose levels. It always worked before. "Oh well" the pancreas would say, "I'll put more insulin into the blood and that will do the trick!"

And the pancreas is right; it does do the trick...for awhile.

Not long after the pancreas decided increasing the amount of insulin a bit is the answer to the 'rebellion', the newly increased amount stops producing the needed result. Again the pancreas says, "I've got this; putting yet more insulin into the blood will do the trick!"

And the pancreas is right; it does do the trick...for awhile.

I'm sure you can see how this ends up going downhill.

This pattern is the first discernible sign of the breakdown of the "daily high blood glucose management system."

Where have we heard the phrase "daily high blood glucose management system" before? Could it be chapter 7 and our discussion of the genetic limits of emergency response system called the Hepatic Lipid System, which encompasses the glucose/insulin cycle?

Perhaps you remember these words; *"Would forcing the body to employ an emergency mechanism - intended to be used perhaps several times a year - to constantly act as your body's 'high glucose management system' produce consequences? If so, what might those consequences be?"* Welcome to the beginning of the consequences.

Insulin resistance is the proverbial "vicious cycle". Over time, as the volume of insulin the pancreas puts out increases in order to "compel compliance" from the cells, the more resistant the cells become to insulin's command.

Let me be clear; this is the first readily discernible sign (absent very specific tests being performed earlier) that your body's ability to use the Hepatic Lipid System as a daily high glucose management system is breaking down. It is beginning to fail.

Insulin resistance is your body's way of clanging the alarm bell loudly to alert you there is a fundamental breakdown occurring that requires your attention, and it produces some notable symptoms.

While there are medical tests available to confirm you have insulin resistance, they are hardly necessary.

As far back as the 1800s British physicians working in the occupied lands of the British Empire noted chronic diseases (previously unknown in the indigenous people) manifested themselves in that portion of the indigenous population that adopted a British (today, "western") diet. Interestingly, the manifestation occurred 20-25 years after indigenous people adopted the British diet. (Indigenous people living outside the main population centers who did not adopt the western diet experienced almost no chronic disease.)

In other words, there appears to be a timeframe after which the body can no longer cope with the abuse of forcing the Hepatic Lipid System to function as a daily high blood glucose management system and begins to deteriorate.

Signs of IS are pretty clear. The most noticeable sign for most people is an increase in body fat that you cannot diet away or exercise away. Body fat just keeps gradually accumulating no matter what you do. You may see a period of brief reprieve here or there, but it won't last and the fat will continue increasing. Depending on your genetic make-up, it may accumulate mainly in the abdominal area, or it may cover a larger area, such as the torso generally, and perhaps the hips and thighs as well.

You will likely feel uncomfortable abdominal bloating after every meal. The further the IS progresses the more aggressive the abdominal bloating and discomfort will become; the worse you will feel after eating.

Another key sign is drowsiness in the afternoon. As the IS gets worse, the number of days you feel drowsy will increase, along with the intensity of the drowsiness.

You will likely get "whatever is going around" more often.

You may see an increase in skin tags.

Any one of these may exist without you having IS. But they will not appear grouped together, and they will not persist, unless it is insulin resistance.

What happens if you ignore your body's clanging warning bell in the form of insulin resistance?

We move from the precursor (IS) to our first chronic disease.

TYPE-2 DIABETES

Left unaddressed, IS advances to type-2 diabetes. In other words, at the beginning of the cellular rebellion the condition is called insulin resistance. As the resistance becomes more pronounced it is called type-2 diabetes.

In some cases, the pancreas works so hard to keep up with insulin demand that it exhausts itself and is no longer able to produce the volume of insulin required to force the cells to comply. Although some may see this as a positive on first blush, it is not. Not only does that leave blood glucose levels dangerously high without a means of resolution, but it is not a proper or healthy reduction in insulin production;

it is a pathology It is another, more serious, level of breakdown in the Hepatic Lipid System based on its unsustainable use as a daily high glucose management system.

Diabetes was virtually unknown in the late 19th century. During that era doctors attended conferences to learn about diabetes. They had to attend conferences because diabetes was not included in medical school curriculum. Most of the attendees had never met a person, no less had a patient, with diabetes.

How different that is from today when 1-in-7 Americans have diabetes, according to the Centers For Disease Control. From the vast majority of doctors never having seen a case of diabetes, to 1-in-7, in 130 years.

Let us now examine another precursor.

SYNDROME X

Here are the 5 constituent elements of Syndrome X again;

- Obesity of the adipose tissue
- High blood pressure
- High level of triglycerides in the blood
- Low HDL
- High blood glucose

One need only manifest 3 of the 5 to be diagnosed as having Syndrome X.

Syndrome X is not a disease, or even a precursor. It is simply a cluster of abnormalities that collectively or individually indicate the body is struggling with something.

Here is a quote from Wikipedia concerning Syndrome X: "The exact mechanisms of the complex pathways of metabolic syndrome [Syndrome X] are under investigation. The pathophysiology is very complex and has been only partially elucidated."

The stupidity of those words is mind-boggling. But...perhaps the author knows those words are utter nonsense. Perhaps it is simply part of the on-going effort to obscure the truth from the public. Perhaps it is yet another example of the propensity to paint as complex that which is not; another avoidance of Occam's Razor.

Here is the simple reality. Consuming a high level of carbohydrates daily, thus forcing the Hepatic Lipid System to constantly act as the body's high blood glucose emergency management system, produces the following physiological results:

- Obesity of the adipose tissue
- High blood pressure
- High level of triglycerides in the blood
- Low HDL
- High blood glucose

Does the list look familiar?

NOTE2: *Blood pressure increases with elevated insulin levels.*

> *"The stronger the body, the more it obeys;*
> *the weaker the body, the more it commands."*
> *~ Jean-Jacques Rousseau (1712–1778)*

I would like to draw your attention to the fact that if we separate those 5 conditions out of the cluster and look at them individually, there is abundant scientific evidence (acknowledged even by establishment researchers) that each is the result of high carb intake. All of them can be quickly and easily reversed by no longer eating any significant quantity of carbs.

Are we to believe that considered individually we know they are driven by the problems from Hepatic Lipid System overuse, but if we consider them collectively - as a cluster - the cause miraculously disappears and somehow becomes a matter too complex for 21st century science?

Not only is that a scam, but it's not even a creative believable scam. It's unadulterated stupidity. Perhaps the worst part is that whoever wrote those words in Wikipedia thinks **you** are that stupid!

CANCER

Cancer is the result of mutation in DNA. There is nothing one can do (at this point in history) to alter the reality of having cancer-enabling DNA mutations, if one has such mutations. As I write these words, science is not even certain how many genes are affected, which ones, or precisely how they are mutated. Lots of speculation. Few hard facts.

Does having cancer-enabling mutation of certain genes mean you will get cancer in your lifetime? Again, because science's understanding is so sketchy, we really don't know.

Will everyone who has cancer-enabling mutations get cancer? Unknown.

Do most people with cancer-enabling mutations actually get the disease? Unknown.

Do few people with cancer-enabling mutations get the disease? Unknown.

You get the point.

While we do not know all we'd like to know about cancer, we do know some rather fascinating things about how cancer cells react to certain stimuli.

Research has proven beyond any doubt that cancerous tumors grow rapidly when exposed to insulin, and do not grow rapidly, if at all, when very little insulin is present (such as at baseline).

Estimates are that cancer cells consume 10 times as much glucose as non-cancer cells.

Imagine the mutated cells are in your body right now. Would you think it wise to saturate cells that may kill you in the very substance - glucose - they crave for energy [energy is life!], which in turn signals the body to produce the hormone - insulin - that dramatically speeds their growth?

A closely related cousin of insulin is "Insulin-like Growth Hormone"

(IGF). Research shows that elevated levels of IGF play a crucial role in spreading [metastasizing] cancer cells from their initial site to other areas of the body.

The kicker here is that high levels of insulin - which of course correspond to high level of blood glucose - signal the body to increase IGF production beyond ordinary levels. (Ordinary = levels in the absence of glucose-driven high insulin.) In other words, higher insulin = higher IGF.

Since probably somewhere in the range of 99.5% of Americans (maybe more) live in glucosis, virtually every person diagnosed with cancer was/is using the emergency Hepatic Lipid System as a daily high blood glucose management system.

To me this seems quite a bit like splashing gasoline on my clothes and then repeatedly jumping over a burning fire pit. How long could that behavior continue without resulting in the inevitable consequence?

By contrast, if one had the mutated genes, yet lived a lifestyle that virtually never produced a significant glucose increase from baseline, and thus never signaled the production of any meaningful increase in insulin, what might happen, or more pointedly, not happen?

We don't know the answer to that.

Why?

Because as we discussed in chapter 10, the nutritional research industry in the United States refuses to fund such…um…"dangerous" studies.

You will remember I mentioned cancer was a rarity before the 20th century.

Do we imagine that cancer-enabling genetic mutations somehow exploded in the last 100 years? Let me assure you that is physiologically impossible. Such a dramatic increase in genetic mutations producing one specific form of disease would require (at a bare minimum) tens of thousands of years.

Yet we have seen an explosion in the number of cancer cases in the U.S. over just a 100-year period. The beginning of that shift, which has produced such a dramatic increase in cancer, coincided with the "Manufactured Food Revolution", allowing virtually unlimited carbohydrate consumption.

But I'm sure that's just coincidence.

A closing note on cancer in the 20th & 21st centuries, unrelated to the glucose/insulin cycle. Researchers have known since the mid-1970s that high consumption of polyunsaturated fats in the form of "vegetable oil" increase the odds of getting colon cancer 300%, and all other forms of cancer by 200%.

> **NOTE3:** *I put "vegetable oil" in quotation marks because it isn't vegetables oil. (Vegetables contain close to zero fatty acids.) What is dishonestly pitched to you as "vegetable oil" is actually industrially processed oil from nuts, seeds, and grains. The extraction of oil from nuts, seeds, and grains only became possible, via industrial technology, at the turn of the 20th Century. Can you imagine any other product concerning which the U.S. Food and Drug Administration and/or the Federal Trade Commission would allow such a blatantly misleading description? But the billion dollar processed food industry doesn't control the actions of federal agencies. Not at all.*

If the trajectory of increase in colon and rectal cancer observed from 1975 to 2010 remains constant moving forward, researchers at the University of Texas MD Anderson Cancer Center tell us to expect a 90% increase in colon cancer in people ages 20 to 34 and a 124% increase in rectal cancer, by 2030.

Despite this, the U.S. government and the medical industry continue telling Americans to get their dietary fat from 'healthy sources', such as "vegetable oil." It's almost as if the government is the taxpayer funded marketing arm of the processed food industry. But that couldn't be true...right?

ALZHEIMER'S

It seems medical and nutritional researchers in the U.S. have a very real problem seeing the nose in front of their face - scientifically speaking.

Imagine researchers were assigned the project of making a test car roll efficiently. The test model with which they are presented has four flat tires.

They spend years trying everything they can think of to make that test car roll efficiently. They perform extensive testing on the car. They consult with vehicle experts. They partner with experts on the subject of movement by the act of rolling. They spend hundreds of millions of dollars in laboratory testing on every conceivable line of scientific inquiry that may lead to a way to make that car roll efficiently.

You are impressed by the level of commitment these researchers have shown to "curing" the "illness" of the car not rolling efficiently. You're aware of their unceasing efforts because it is so often reported in the press.

You happen to be visiting the community where their laboratory is located and decide to stop by and let the valiant researchers know how much the nation appreciated their ceaseless efforts to find a cure for the scourge of cars not rolling efficiently.

As you're being escorted to the office of the intrepid researchers you pass through the lab. There you see the test car. What a grand moment!

But wait…. As you look at the test car you find it hard to believe your eyes. This simply cannot be true. The car still has four flat tires!

The researchers, being so intent on "discovering" a way to make that car roll efficiently, looked right past the common sense measure of inflating the tires. After all, anyone can inflate the tires. Where's the glory in that?

That tongue-in-cheek tale speaks to how researchers all-too-often see their job, and the public sees researchers. The public sees researchers as heroes arduously battling the scourge of disease, and researchers see their job as finding a solution - just as long as the solution isn't fairly obvious and doesn't involve common sense.

Alzheimer's has been the subject of decades of intensive research at the cost of hundreds of millions of dollars. Researchers will tell you they've made no progress in preventing Alzheimer's or finding a cure. Neither is considered likely absent some unforeseen breakthrough.

What we do know is that when a person dies from Alzheimer's and researchers examine the brain, they find there something referred to as "amyloid lesions" (also called "amyloid plaques"). Amyloid lesions are not present in the brains of people without Alzheimer's. These lesions are considered a leading factor in the onset of Alzheimer's.

The lesions are called "amyloid lesions" because they involve "amyloid beta", an amino acid (protein) peptide present in the brain (and elsewhere). Science does not currently understand the role of amyloid beta in the body.

Nevertheless, it appears that in the brains of those with Alzheimer's, amyloid beta peptides cluster together at numerous sites, thus creating the "plaques", or "lesions". The plaques/lesions disrupt proper neuron function in the brain, producing the affects we recognize as Alzheimer's.

Yet there are other things we know.

Healthy brain tissue and high insulin levels are incompatible. The brain protects itself from high insulin levels by producing an enzyme called "Insulin Degrading Enzyme" (IDE). IDE does precisely as the name implies; it degrades the chemical composition of insulin so the insulin can no longer act as insulin. That process was first observed in the mid-20th century.

But IDE has a secondary role. It breaks down amyloid beta. Breaking it down allows the amyloid beta to be cleared from the brain.

The body has natural limits to the production of every substance it creates. IDE is no exception. In other words, only so much IDE is available for use in the brain. There is no 'volume knob' you can turn to increase IDE in the brain.

In the brains of our ancient ancestors, who existed in ketosis, incidents of high insulin would have been incredibly rare. That left IDE in the brain free to degrade amyloid beta.

But that simple healthy process isn't how things work today, is it?

Today Americans create high insulin levels multiple times a day, every day.

Because insulin travels right through the brain's blood barrier, each and every time a person spikes insulin during the day, the high insulin level is experienced by the brain. IDE leaps into action, degrading the insulin to protect the brain. The enzyme itself (IDE) is consumed during the process of degrading the insulin. The higher the insulin level, and the more often it is elevated, the more IDE is "burned up" countering the problem.

After combatting high insulin all day, how much IDE is left for other tasks, such as clearing amyloid beta from the brain? We don't know.

We don't know because there is no recognition or acceptance by the research industry that high insulin levels play a key role in Alzheimer's. No research has been done that would answer that question for us. And none is planned.

Absent research that would tell us how much IDE is left for amyloid beta clearance after being consumed by repeated bouts against high insulin throughout day, we are left to hypothesize.

That brings us back to our body's hardwired genetic coding, which is a constant, and foundational to every aspect of our bodies.

If we agree our ancient ancestors lived for millions of years in ketosis, and thus the systems of our body are tailored to act within that construct, can you see any reason the body would have developed in such a way as to produce enough IDE to battle high insulin in the brain all day long and still have enough left over to clear amyloid beta?

In a multi-million year environment in which high insulin levels were virtually unknown, there would be no purpose for the body to develop the capacity for high output of IDE. Since there was to no purpose for the body to develop that capacity, it is eminently reasonable to hold the view that it did not develop that capacity. There is no science indicating it does.

This leaves a logical person with the construct that the body creates sufficient IDE in the brain to tackle small occasional upticks in blood insulin levels, clear amyloid beta, and perform its other (currently unidentified) functions. But...the body does not produce enough IDE to battle high brain insulin all day long **and** effectively clear amyloid beta from the brain. In that environment there simply isn't enough IDE to go around. Queue Alzheimer's.

I believe it is important to state plainly that nutritional and medical researchers are 100% uninterested in the question of whether people living in ketosis never get Alzheimer's.

I can almost hear their response if medical researchers were told people in ketosis never get Alzheimer's; "That's not the focus of the research we've been funded to perform."

Right. Wouldn't want to inflate those tires.

OBESITY

Based on what you've learned thus far I'm confident you can connect the dots yourself. If not, may I suggest re-reading chapter 6.

For the sake of completeness I will address some basics.

> *"If obesity is a disorder of excess fat accumulation,*
> *what regulates fat accumulation?"*
> *~ Gary Taubes, Science Journalist*

Let me begin with this simple statement: There are no obese people who do not consume a lot of carbs.

High carb intake -> high monosaccharides -> high blood glucose -> high insulin. High blood glucose engages the HLS (chapter 6).

At the same time the liver is pumping triglycerides out in VLDL particles, insulin is also signaling adipose cells not to mobilize fatty acids. In other words, here come the VLDL particles to deposit more fatty acids into the adipose cells while insulin demands the cells hang onto what they already have.

The high glucose level, and corresponding high insulin level, also ensures all 100 trillion cells become chock full of glucose. That means the cells have little need for energy from the triglycerides. Accordingly, the triglycerides either dwell in the adipose cells or remain in the blood, which is reflected in a high triglyceride count on a blood test. (As Dave Feldman poignantly observed, there is no "engineering reason" for energy to be stored in the blood.)

When a body is functioning in glucosis, cells oxidize glucose for energy and start to run low. A feedback mechanism signals the brain to replenish the glucose. This is when you feel hungry.

The cells could burn fatty acids stored in the adipose tissue, but that would require a delay in feeding, with insulin sitting at baseline for some time. The vast majority of Americans never get there. Long before the body would begin mobilizing fatty acids from the adipose cells the person eats his/her next meal.

That meal, like those before, will be high in carbs. Repeat as necessary to gain additional body fat. Add a little insulin resistance (which begins years before it becomes a discernible problem) and the adipose cells just keep expanding. As the process continues people gain weight. Several years later they are noticeably overweight. Several more years and they have become obese.

Perhaps there is a reason 80% of Americans currently fall into the categories of either "overweight" or "obese", clinically speaking.

As I close on obesity I wish to state plainly, it is **all about the insulin.**

We know this because when we take an overweight person and alter his/her macro intake in such a way as to keep insulin shallow and stable, their excess body fat falls away effortlessly - even when eating considerably more calories than on their previous high insulin diet. (Science has known since the early 60s that elevated insulin demands storage of fatty acids in the adipose tissue.)

Let me quickly address the common assertion that obesity is genetic; hereditary to some extent. Let us recall that 100% of research on such matters has been performed on a glucosis society.

I will not here delve into obesity research (such as it is). Instead I will suggest you view such a genetic disposition in the following light: What is being observed is not a genetic disposition to obesity, per se, but rather that person's body simply cannot withstand the repetitive physical abuse of the vicious glucose/insulin cycle to the same extent as can others. The result being early and/or excessive weight gain.

HEART DISEASE

Atherosclerosis is the single biggest health concern in our culture.

It is also the issue that was the springboard for the Ancel Keys' falsified data - combined with his formidable political skills - convincing an entire nation of the falsehood that dietary fat, particularly sat-fat, causes atherosclerosis and heart attacks.

The party line from the research and medical industries is they don't know what causes heart disease. As the American Heart Association puts it…"Exactly how atherosclerosis begins or what causes it isn't known."

NOTE4: *A sure way to get chronic disease and die earlier than you otherwise would, is to follow any advice from the American Heart Association, the American Diabetes Association, or Harvard Public Health. These organizations are, in my view, key players in protecting the establishment narrative.*

NOTE5: *In November 2017, the President of the American Heart Association (who is a cardiologist) had a heart attack - at 52 years of age.*

What causes atherosclerosis is a highly detailed topic.

Note I say "detailed" rather than complex. Like other subjects I have addressed with you, the core reality is not complex, but there are so many intricate details involved in the subject that discerning which of the various factors are the 'lynchpin' issues is indeed challenging.

Keeping with the theme of this book, I'm going to distill the issues down to the key points and not spend pages upon pages detailing the underlying science. (As always, taking what you learn here and doing your own in-depth research is encouraged.)

There are numerous factors that create the overall environment in the arteries of the heart leading to myocardial infarction, commonly known as "heart attack".

Every single one results from forcing the body to use the Hepatic Lipid System as a daily high-glucose management system. Period.

If a person continues forcing his/her body to use the Hepatic Lipid System as a daily high-glucose management system, that person **will** (not "might") create the necessary conditions that significantly increase the odds of myocardial infarction.

1-in-4 Americans die from heart disease and it is the #1 cause of death in the U.S. Medical researchers claim they still don't know why Americans have such an incredibly high rate of heart disease.

I am 100% certain those at the top of the research power structure do know. If they did not, how would they know which type of research is never to receive funding?

The medical industry does offer a list of things one can do to (allegedly) prevent heart disease. The list is a combination of common sense things and things long-since disproven by science.

NOTE6: *Science doesn't actual 'disprove' things. All science can do is prove a thing factual. However, if researchers have attempted for several decades, in numerous studies of differing types, to prove a hypothesis factual - at the cost of hundreds of millions of dollars - and have been unable to do so, it is a virtual certainty the hypothesis is not factual/true, hence my use of "disproven".*

The common sense suggestions are things such as 'don't smoke' and 'keep your triglycerides low'.

They also offer advice for which zero clinical evidence exists, such as "Try to limit saturated fats, foods high in sodium...". Neither sat-fat nor sodium have anything to do with the onset of atherosclerosis. Yet the medical industry continues spewing this nonsense. (In a little bit I'll explain why they continue spreading harmful myths to the public.)

They offer this advice: "Eat plenty of fresh fruit, vegetables, and whole grains." Uh...no.

"Eat plenty of fresh fruit" is in direct conflict with one of their other

items of advice; "Stay at a healthy weight." I know you've been told your entire life that fruit is good and healthy. Do you remember the list of non-factual "nutritional myths" in chapter 1? Do you remember the question I posed to you? *"Can you honestly say your belief in any of the above statements is the result of you reading clinical study reports?"*

If you understood the chemistry of how fruit - actually, the fructose in the fruit - reacts in the body you would understand that an occasional small piece of fruit for your enjoyment is fine, but there is zero health benefit from eating fruit, and eaten in quantity fruit is injurious to your health.

Yet the medical industry tells you to eat "plenty" of it.

They recommend that you "Eat plenty of...vegetables..."

What have we learned about carbohydrates?

Do you recall this from chapter 6, *"All carbohydrates - whether table sugar or broccoli - are broken down in the intestines to monosaccharides. That is an absolute. There is no exception. It's a simple, singular, and linear process; carbs -> monosaccharides"*?

Yet the medical industry tells you to eat "plenty" of them.

NOTE7: *Consumption of small amounts of low-carb vegetables is not problematic.*

"Eat plenty of...whole grains." <sigh> Grains are nearly 100% carbs. And starchy carbs at that, which are arguably the worst form of carbs, excepting of course sugars.

Isn't it great that the medical industry continues sharing advice with you that was debunked long ago, and suggests you eat "plenty" of the very substances that actually contribute to heart disease!

Then they offer the tired old line about keeping cholesterol low to prevent heart disease. What's next - bathing is the cause of illness? Maybe a return to blood letting!

The subject of cholesterol could fill an entire chapter by itself. In this book I will only mention the following:

Roughly 70% of your cholesterol is manufactured by your own body, in virtually every cell and in the liver.

Cholesterol is, in reality, a protective substance, not a harmful one.

Your body knows exactly how much cholesterol it needs. The amount the body determines it needs is correct and healthy, not unhealthy.

If you eat too little dietary cholesterol your body will simply produce more of its own. If you eat a high level of dietary cholesterol your body will compensate by making less of its own. It knows the amount it needs to be healthy.

Like virtually all medical industry "norms", "healthy" levels of cholesterol have been set for people living in glucosis, not ketosis. Even then they have it wrong.

What the medical industry considers "high" cholesterol is not an indicator of...anything...unless it is part of a cluster, such as Syndrome X.

Let me be clear about this. If your triglycerides are under 100 and your HDL is above 50, cholesterol (usually shown on blood tests as "LDL-cholesterol") running all the way into the 400s is not indicative of any increase risk of atherosclerosis.

So why does the medical/pharmaceutical complex keep preaching that everyone must lower their cholesterol? Perhaps it has something to do with the fact that sales of cholesterol-lowering drugs are projected to hit $1 trillion in 2020. This despite the fact that there is no clinical evidence showing cholesterol causes heart disease.

> *Medical science has made such tremendous*
> *progress that there is hardly a healthy human left.*
> *~ Aldous Huxley*

I don't want to turn this into a cholesterol chapter, so I'll simply leave you with the fact that low cholesterol is the real problem because of cholesterol's heart-protective qualities.

Did I mention all 5 of the oldest people in America (over 100 years old) have an LDL-cholesterol level of 130 and above. (The medical industry says LDL-C should be under 100; a number unsupported by scientific evidence.) Further, 4 of the 5 have LDL-C above 165!

Did I also mention longitudinal data from the Framingham study shows low cholesterol is associated with higher "all-cause mortality" than either "normal" or "high" cholesterol (as such terms are defined by the medical industry)?

NOTE8: *Framingham is one of the most notable large-scale, incredibly expensive, nutritional/heath research studies ever conducted, with new data sets being examined decades after the primary study period.*

> *We have a pharmacy inside us that is absolutely exquisite.*
> *It makes the right medicine, for the precise time,*
> *for the right target organ - with no side effects.*
> *~Deepak Chopra*

As I move on from cholesterol, let me leave you with a broad statement about all substances produced by the human body:

The idea that the body produces substances that kill itself, absent unhealthy influences by the conscious occupant, is absurd!

I could go on and on with the ridiculous advice provided to the American people by the medical industry under the guise of helping you prevent heart disease, but you get the point.

So what does cause heart disease?

I'm confident by now you're pretty sure of the answer; forcing the body to use the Hepatic Lipid System as its daily high glucose management system.

Because heart disease remains such a concern for many Americans, I'm going to provide a deeper level of information concerning how the daily abuse of Hepatic Lipid System creates atherosclerosis.

In the HLS chapter you learned that the liver converts excess blood glucose to triglycerides, packages them up as VLDL particles and sends them out into to the blood, where they are stored in adipose cells.

Let's now look at timing.

When a person who has not yet developed unhealthy insulin patterns eats a high-carb meal, insulin is fairly close to being back to baseline at the 2 hour mark. This means every task insulin demands the various mechanisms of the HLS to perform must be accomplished within that time.

For the purpose of examining heart disease we are focusing on liver synthesis of glucose into triglycerides, and the VLDL particles by which those triglycerides are released into the bloodstream.

For roughly an hour and half of the aforementioned 2-hour window the liver is frantically converting glucose to triglycerides and releasing them into the blood in VLDL particles. VLDL particles transfer their triglycerides to the adipose cells and are then downgrade from VLDL particles to LDL particles.

NOTE9: *In ordinary (non-emergency) creation/release of VLDL particles, VLDL contains a good deal of cholesterol in addition to triglycerides. In that scenario VLDL downgrades first to IDL, downgrading to LDL only after off-loading the cholesterol. However, when VLDL is created on an emergency basis - the HLS - there is very little cholesterol present and so the downgrade moves directly from VLDL to LDL, skipping the IDL phase.*

LDL Particle Count:

Research shows an association between a high LDL particle count (identified in blood tests as "LDL-P") and heart disease. This makes perfect sense.

Earlier I said we'd be looking at fatty acid mobilization during fasting. That time has arrived. But before we look at fatty acid mobilization while fasting, we should take a moment to review how triglycerides are introduced in the body.

Method 1) You consume dietary fat and those fatty acids (in the form of triglycerides) are introduced into the body via the Lymphatic Lipid System.

Method 2) The liver converts excess blood glucose into triglycerides via the Hepatic Lipid System.

I refer to the LLS & HPS as "systems" because the fatty acids did not exist in the body until introduced by one of these two pathways, i.e. 'systems'.

In contrast, the fatty acids mobilized during fasting already exist in the body, having previously been stored in adipose cells.

As we all know, eating provides the fuel for the 100 trillion cells of the body. And of course if we don't eat - we don't give the body fuel - eventually we die.

There is a state in between eating regularly and dying from a lack of fuel. We call it fasting.

When a person fasts there is no dietary source of fuel so the body turns inward for its fuel; something it can use for cellular energy.

The very first place the body looks is to the fatty acids stored in adipose tissue. This makes perfect sense in terms of our body's hardwired genetic coding because the mitochondria of our ancient ancestors oxidized triglycerides (and ketones) for energy.

NOTE10: *The liver synthesizes ketones from fatty acids.*

When fasting, glucagon (another hormone produced by the pancreas) signals enzymes to initiate mobilization of the fatty acids in the adipose cells. The ensuing enzymatic process alters the condition of the fatty acids, preparing them to exit the adipose cell and be transported.

You will recall that fatty acids are hydrophobic, which is why they are packaged in chylomicrons (LLS) or VLDL particles (HLS) to move through the blood.

There is yet another transport mechanism for fatty acids. (Isn't physiology fun!)

Fatty acids released from adipose cells are picked up and transported through the blood by a protein called albumin.

Albumin carries the fatty acids back to the liver where they are grouped into triglycerides and packaged up as VLDL particles.

I say "back" to the liver because they were originally synthesized in the liver, from excess blood glucose, and stored in the adipose cells as part of the HLS (for those in glucosis).

You may also recall that the liver cannot retain triglycerides and remain healthy, so as soon as the newly arrived fatty acids are packaged up as VLDL particles the liver puts them into the bloodstream.

From here the process is routine. The VLDL particles off-load triglycerides, become LDL particles, and arrive back at the liver where they cease to exist.

In regard to understanding the cause of heart disease, the most important factor in the use of fatty acids during the state of fasting is that the process of mobilization from the adipose cells is gradual; unhurried.

The body sees the release of stored fatty acid from adipose cells during fasting as normal and routine. There is no urgency (though, as we'll discuss shortly, long-term fasting operates a bit different).

The body uses several feedback mechanisms to determine the amount of energy the cells require and begins a steady process of meeting those needs through the gradual release of stored fatty acids.

The process continues - at a steady pace - around the clock. As such, the volume of VLDL particles being put out in any particular period of time is significantly less than when the HLS is operative.

In the following illustrative example I'm going to use artificially low numbers to clarify the concept I expressed in the above paragraph. (The concept might be difficult to grasp if I was speaking in terms of many quadrillions.)

During the 1.5 hour window the HLS system is commanding the liver get rid of the excess blood glucose, the liver will put out 100,000 VLDL particles.

By contrast, during fasting the liver will put out 300,000 VLDL over 24 hours.

HLS = 100,000 VLDL released into the bloodstream in 1.5 hours.

Fasting = 300,000 VLDL released into the bloodstream over 24 hours.

If we break that down per-hour, it's 66,666 per hour in HLS, and 12,500 per hour when fasting.

At this point I need to let you know that when an LDL particle is terminated in the liver, the medical industry refers to that as being "cleared". How many LDL particles are terminated is referred to as the "clearance rate".

As we have discussed several times in various contexts, the body does what it is genetically hardwired to do - at the speed, or in the volume, its genetic coding dictates - and within the limits the genetic code has established.

As an example, we know the HLS is an emergency mechanism meant to defuse a dangerous situation [high blood glucose] quickly. The body is genetically coded to resolve that situation quickly because high blood glucose is toxic to organs and tissue. Phrased simply; danger = speed of resolution.

There is zero danger from fasting. In fact, fasting is a healthy state with quite a number of physiological benefits.

Because there is zero danger to the body associated with fasting, and brief periods without food certainly would have been the case for our ancient ancestors, our hardwired genetic coding for the process the body uses to address fasting takes place at a measured pace, determined by the body's own determination of how much energy the cells require.

The result of this measured pace of VLDL output is a comparably low number of VLDL particles in the blood at any given time.

We should now take stock of, and compare, the difference between the measured pace of VLDL released during fasting with the frantic pace of VLDL released during the emergency HLS response.

The difference in the speed with which the VLDL particles are put out by the liver translated into a differing volume of VLDL in the blood during that period of time.

The volume/time equation has a bearing on the clearance rate. And the clearance rate has a bearing on atherosclerosis!

Remember a moment ago I said, "...*the body does what it is genetically hardwired to do - at the speed, or in the volume, its genetic coding dictates...*"?

This rule also applies to the speed at which the liver clears LDL particles.

I have read seemingly endless research on LDL clearance rates. Two important facts stand out.

1. There are certain physiological circumstances that reduce the LDL clearance rate.
2. There is no known physiological circumstance that increases the clearance rate.

In other words, the body's hardwired genetic coding sets a rate - a pace - of LDL clearance by the LDL receptor sites in the liver. The only known alteration of that rate is downward.

> **NOTE11:** *LDL particle clearance is also performed by cells outside the liver, but the liver accounts for (roughly) 90% of all LDL particle clearance. Evidence indicates the non-hepatic LDL clearance rate always remains constant, again confirming the body functions at the pace fixed by its hardwired genetic coding.*

Now that we understand the liver has a genetically set LDL clearance rate, let's look at how that reality impacts whether or not a person gets heart disease.

Unsurprisingly, that takes us right back the HLS.

We have already established that, based on our genetic coding, the HLS is an emergency mechanism intended to clear a toxin - excess glucose - from the blood on the rare occasion ancient man found enough carb-rich food to overindulge in it.

In attending to that task the liver frantically converts glucose to triglycerides and puts them into the bloodstream in VLDL particles.

In other words, we have an emergency mechanism putting out an exceptionally high volume of VLDL particles, while at the same time no mechanism exists to increase the liver's clearance rate correspondingly.

Let's use a non-physiological example to clarify the point.

You have the ability to read one 400-page book a day. No more.

You're taking a class at the local college and the curriculum requires you read a single 300-page book each day. No sweat.

As the class progresses the requirement changes to two 200-page books a day. Doable!

The requirement changes again, to two 300-page books a day. Uh oh! At your maximum reading pace you can only complete 400 pages a day. That limit, and the new course requirement, means there are now 200 pages each day you can't complete. The demand now exceeds your capacity.

That is precisely the situation in which the human body finds itself when the HLS pumps out more VLDL particles than the liver can clear on a timely basis. For most Americans that's multiple times a day, every day!

What is the consequence of the HLS producing more lipoprotein particles than can be timely cleared?

If you were ancient man, whose body experienced an overabundance of VLDL particles perhaps a couple of times a year, nothing.

For the average American, whose body is producing an overabundance of VLDL particles 3, 4, 5, or 6 (or more) times a day, the consequence is... profound.

"Familial hypercholesterolemia" is genetic disorder in which the ability of the body to remove LDL from the blood is impaired.

Though the genetic mutation may involve the LDL receptors, or ApoB100 protein (or other factors to a lesser extent), the mathematical reality is always the same; a significant increase in the total number of LDL particles and the particles remaining in circulation longer than normal.

Phrased simply, the genetic impairment results in too many LDL particles being in the blood for too long.

But, must one have a genetic disorder to have too many LDL particles in the blood too long?

If one has even a passing familiarity with mathematics, and applies a wee bit of logic, the answer is clearly "No!"

We know the body is genetically coded to consume virtually no carbs, which means the body's natural state in very low VLDL production. In comparison to how Americans eat today, we might say that in its proper genetically coded operation the body produces **incredibly few** VLDL particles.

As discussed earlier, VLDL degrade to IDL, then LDL. Our ancient ancestors would have had almost no LDL in their blood because they started with very few VLDL.

Accordingly, there was never a need for the liver to develop the ability to clear a high number of LDL particles in a short time.

Further, because the HLS was engaged so rarely in ancient man, there would have been no need for the liver to develop a mechanism by which to increase its clearance rate.

What we end up with is genetic code that is the result of millions of years of scant VLDL production, coupled with a fixed rate of LDL clearance, which cannot be increased.

All of that works in perfect harmony for a body in ketosis. But how does it work the way most Americans eat?

Let's again use some artificially low numbers to illustrate how hundreds of millions of Americans create their own self-imposed familial hypercholesterolemia-type condition **every day.**

For the purposes of this example, let's say your body can clear a maximum of 10,000 LDL particles in a 4-hour period.

A person eats a high carbohydrate meal, thus engaging the HLS.

Thus engaged, the liver releases 35,000 VLDL particles into the blood.

Through various mechanisms too complex and lengthy to go into here, the person ends up a short while later with 25,000 LDL particles remaining in the blood.

The liver can clear 10,000 every 4 hours, but there are 25,000 circulating. But wait! We're not done.

Three hours later the person eats again. To keep this illustration simple, we'll say he/she eats the same meal, which means he/she just created another 25,000 LDL particles.

Just before the second meal was eaten, the liver had cleared 7,500 of the original 25,000 LDL particles, leaving 17,500 still circulating.

The new meal adds 25,000 additional LDL particles to the mix, thus increasing the LDL particle count to 42,500 - with a max clearance rate of 10,000 per 4 hours.

Four hours later the person's LDL particle count is down to 32,500. Then the person eats again. (To keep this illustration simple, he/she ate the same meal.) Now his/her LDL count is 57,500!

This illustration presents the exact same outcome as from familial hypercholesterolemia - i.e. a higher and higher number of LDL particles circulating in the blood, created by dietary choice rather than a genetic mutation.

Did I mention it is a virtual certainty that people with familial hypercholesterolemia get heart disease?

So...if by diet a person creates the exact same LDL situation as does familial hypercholesterolemia, what would make that person think his/her outcome will be any different from actually having familial hypercholesterolemia?

NOTE12: *Even though the name of the condition is familial hypercholesterolemia, it is solely the LDL particle count that is relevant to this discussion.*

We know statistically people with a consistently higher LDL particle count are more likely to get heart disease. Yet with its blinders on, the medical and nutritional research industries cannot put 2 and 2 together concerning the circumstances that lead to a high LDL particle count.

Before we move on I want to take a moment to mention 'extended fasting' because it has a different dynamic than the commonplace short-term fasting we've been discussing. (I define 'extended fasting' here as more than 24 hours.)

Fasting for longer than a day produces an increase in the LDL-P count.

Because so few people engage in extended fasting I'm not going to take your time to discuss the reasons for its unique physiological effects. Instead I will simply observe that extended fasting is - and should be - something that takes place rarely (if it all).

NOTE13: *Dave Feldman's data indicates the leaner and/or more fit/athletic one is, the greater one's odds are of having elevated LDL-C and LDL-P, when in ketosis. As Feldman points out, absent inflammation - which is essentially unheard of in ketosis - there is no evidence these elevated readings indicate any health risk. I direct your recollection to my earlier statement that the body does not produce substances that harm itself, or produce them in quantities that harm itself (absent some unhealthy interference from the host consciousness).*

LDL Particle Size:
Another factor bearing upon risk of heart disease is LDL particle size. The two descriptors for particle size are:
'Pattern A' means "large fluffy" LDL particles.
'Pattern B' means "small dense" LDL particles.
Pattern B is an indicator of increased risk.
Pattern A, low risk (if at all).

It should come as no surprise then that a high LDL particle count (increased risk) is associated with Pattern B' (increased risk).

A lower LDL particle count is associated with Pattern A.

There is also a positive association for pattern A size with low triglycerides and high HDL. Again, no surprise.

Since pattern A - which is what we want - is associated with low triglycerides, we might ask what increases triglycerides, because we don't want to go there, right?

High carbohydrate intake is the cause of elevated triglycerides. (A fact universally acknowledged, even by establishment researchers). As you'd expect, low carbohydrate intake lowers blood triglyceride levels.

Have you noticed how everything healthful for the body dovetails perfectly, one with the other, from the perspective of it being ketosis?

NOTICE: **For those with an interest in the subject of type-1 diabetes, after the final chapter you will find an appendix item detailing my hypothesis on how to prevent the onset of type-1 diabetes.**

To keep this book short, the subject of how to place your body into ketosis, and continue "living in ketosis", is not addressed herein. If you'd like to receive (at no cost) a paper I wrote entitled "Getting Started In Ketosis", simply go to drreality.news/reports and input your email address. The document will automatically be sent to you. (Your contact information will always remain confidential.)

CHAPTER TWELVE

THIS CHANGES EVERYTHING!

"All truths are easy to understand once they are discovered; the point is to discover them."
~ Galileo Galilei

I hope this book has provided you valuable information and presented you with a new and healthier way of looking at nutrition, diet, physiology, and chronic disease.

But there is one more thing we need to discuss.

It is the 'nuclear bomb' I've been holding back.

In this book you learned there are only two modes in which all human bodies operate in terms of what substance the cells oxidize for energy. A body can be in glucosis or ketosis, oxidizing glucose or fatty acids & ketones, respectively.

In chapter 6 I revealed to you that the HLS is the body's emergency response system to cope with the rare overconsumption of carbs by our ancient ancestors.

You may have gotten to this point thinking that is the body's only emergency mechanism for removing excess blood glucose. It isn't.

There is another.

The primary emergency system the body employs to get rid of toxic high blood sugar is GLUCOSIS.

Yes, you read that right.

In our modern world society sees glucosis and ketosis as equal options;

pick one; it doesn't matter which you choose; either is fine.

Like virtually everything else we've been programmed to believe, that simply isn't so.

They are not "equal". One is not "as good as" the other.

The fact of the matter is one is normal and healthy for the body, acting in harmony with the body's genetic coding.

The other is not merely one of two acceptable methods of fueling the cells, but is actually the body's emergency response system to rid itself of toxic excess blood glucose.

When blood glucose rises, that demands an insulin response, right?.

So far we've seen that insulin tells the liver to frantically begin converting blood glucose into triglycerides in order to rid the body of the toxic high blood glucose.

That is an hours-long process leaving a toxic elevated level of glucose in the blood as the liver struggles to overcome the problem.

What additional measure can the body take to hasten the removal of the excess glucose?

Simple. Shove it into the cells and burn it away!

Phrased another way, the body employs two mechanisms working together as one comprehensive system to rid the body of glucose in the blood above baseline:

1. Liver synthesis of glucose into triglycerides (which cells in ketosis are genetically coded to burn as fuel).
2. Combustion; burn it.

Glucosis is not merely one of two acceptable modes in which to operate the body. Glucosis is a crucial part of the body's emergency method of getting rid of toxic levels of blood glucose.

I have referred to high levels of blood glucose repeatedly as "toxic". That's not something we hear in this modern glucosis-centric world. Yet it is factual.

Health problems acknowledged by the medical industry as a result of unchecked high blood glucose include, heart disease, cognitive decline, bone & joint problems, kidney disease, sexual disfunction, depression, stroke, nerve damage, vaginal and skin diseases, eye damage, ear infections, mouth infections, urinary tract infections, respiratory infections, yeast infections, and stomach & intestinal problems.

Any question that high blood sugar is toxic to the organs and tissues of the human body?

Any question why the body goes into an emergency mode when high glucose is detected?

Any question why, because the HLS cannot alone cope with the scale of the problem, the body's additional mechanism is to place the toxic material in containers (the cells) and burn it?

Imagine your backyard was filled with piles of trash and you had an incinerator. What would your solution be to get rid of the trash?

You might send some items to be recycled into something useful [analogous to glucose being converted to triglycerides]. But beyond that, the answer is obvious - burn it.

As we've discussed, and hopefully this book has made abundantly clear, the body is intuitively brilliant. If it has the ability to eliminate a toxin by burning it, it would do so - and it does!

"The least questioned assumptions are often the most questionable."
~ Paul Broca

Some people may be wondering how glucose can be so prolific in nature and yet toxic to humans when elevated above baseline. We often read nonsense statements such as, "Glucose is a primary source of energy for living organisms." There is no scientific support for such statements.

Glucose does provide important functions in nature, such as supporting the formation of cell walls in plants.

The World Health Organization refers to glucose accurately when it lists it in medical literature as "intravenous **sugar** solution"! [Bold added by this author.] And what have we learned about sugar's affects on the body?

At baseline glucose is safe for the body. When the body is in a state of ketosis glucose remains at baseline. Exactly where nature intended it to stay; the level at which your body remains healthy.

When glucose levels rise above baseline, as we just saw in the list of damage that occurs to the body, it is dangerous. Some people will correctly state that damage comes from prolonged levels of high blood glucose. And their point is…?

The fact that high blood glucose creates that damage is incontrovertible evidence that glucose is toxic and should never be elevated. The fact that serious damage requiring medical intervention is not done (or perceived) until the body can't cope anymore and starts breaking down from the abuse, hardly refutes glucose's toxicity. It substantiates it.

Today Americans drive glucose up, and then back down. And then up, and down. And up again, and down. And up, and down. In a single day!

The fact that the body gets a few respites from that abuse hardly ameliorates the problem, or the impending consequences.

Stop here for a moment and look at the list of damage again. Note that it spans virtually every form of tissue, critical organs, and almost every functional system in the body.

Why is that? Why are its affects so wide-spread? Because high glucose is **toxic** to **every cell** in the body!

Some folks may be thinking, "Now wait… How can a body in glucosis use glucose as its primary energy source **and** that energy source is also toxic to the body?"

The answer is simple. The body hasn't "chosen" any of this! In our modern society (meaning a few thousand years in this context) the body has been given no option but to adopt the toxin as its primary source of energy. As long as blood glucose continues being elevated, the body **must** reduce it by the two means discussed; HLS and having the cells incinerate it.

Earlier I mentioned that when glucose levels in the cells run low the body's feedback cycle produces the physiological sensation we recognize as hunger. I did not say the cells want more glucose. They don't!

They are genetically coded to use triglycerides and ketones for energy.

Sadly, the on-board consciousness making the decisions just keeps shoving carbs into the body. And the body keeps doing what it must to survive. Then people say things like, "The body's preferred energy source is glucose."

Everyone living in ketosis and experiencing the tremendous life-changing benefits therefrom knows the body's preferred energy source most certainly is **not** glucose. The only people making that silly claim are those living in glucosis (who generally are unwilling to make the shift to becoming healthy).

> *"Those who cannot change their minds cannot change anything."*
> *~ George Bernard Shaw*

A quick word about "carb addiction". I don't want to veer into the complex issue of addiction so I'll simply make two statements.

1. Anytime the body is addicted to any substance it is indisputable evidence the substance is not healthy for the body.
2. If one shows the inner fortitude to resist carb cravings for the 3-4 days it takes for the body to convert to ketosis, the cravings drop to almost zero.

In order to hold this book to a reasonable length I have focussed only on chronic disease, and only those that are of significant concern to the public. Yet many other conditions/diseases that fit the definition of "chronic disease", but are not usually discussed as such, are also increasing rapidly to levels never before seen.

Why is that?

Because every time glucose is elevated above baseline it stresses - initiates inflammation in - every cell in the body a little bit more by exposing them, yet again, to a toxic level of glucose.

Imagine if you ate a bit of arsenic with every meal. That is analogous to what people do to every cell in their bodies via daily bombardment of high glucose.

Taking it a step further, the stress/inflammation to the cells is increased significantly when, as part of the emergency process of eliminating the toxic substance, the only choice the cells have is to bring the poison inside themselves so it can be incinerated.

One of the facts the establishment doesn't want you to know/understand is the very act of oxidizing glucose in the cells' mitochondria creates a byproduct known as Reactive Oxygen Species (ROS), which is a primary cause of inflammation!

For reasons that should be readily apparent to you by this point, establishment minions assert this process is "healthy and normal", while at the same time acknowledging a "significant amount" of ROS produces oxidative stress.

"A significant amount...". Such as every one of the 100 trillion cells in the body constantly producing ROS, all day long, every day! (Of course they assert it's "healthy & normal". According to them, the entire process that makes Americans sick and kills them is "healthy & normal".)

Every cell in the body is put through this several times a day - every day - year after year - decade after decade. And then we wonder why our nation is in the worst health crisis mankind has ever known. And the "experts" tell us this is all just...normal.

Virtually everyone who has studied physiology - including even most establishment minions - agrees the underlying cause of countless illnesses and unhealthy conditions is inflammation. Yet theories about the sources/causes of the inflammation are widely divergent. That makes sense because everyone has been trying to explain something not evident to them; high glucose causes inflammation in every cell of the body; a bit more every day.

Let me make a very clear statement: Of every living entity on Earth that uses glucose in some manner, **ONLY man elevates blood glucose from baseline.** And that only started several thousands years ago when man unknowingly misused technology to separate himself from the Earth/Mankind symbiosis that kept him healthy.

Even ruminants, which eat exclusively plants in the wild, uptake virtually no glucose from what they eat. Just like man in ketosis, ruminants convert fatty acids (absorbed from rumen by bacterial fermentation) into glucose to maintain baseline.

Did I mention that virtually everyone with clinical knowledge of ketosis agrees it is the most anti-inflammatory state in which the body can exist?

> *"Our own physical body possesses a wisdom which we who inhabit the body lack. We give it orders which make no sense."*
> ~ Henry Miller

Insulin resistance - the body's shrieking alarm - is simply the cells saying "ENOUGH! We can't do this any more!"

The medical industry calls it 'insulin' resistance because in this glucosis-centric world it's unfathomable - beyond the pale to even consider - that what is really being resisted by the cells is the absorption of any more toxic glucose.

Because I am the first person to declare glucosis a part of the body's emergency glucose clearance system - and not merely an "acceptable choice" - there will be those who refuse to accept it.

Some people will have difficulty coming to terms with the fact that they, their parents, their grandparents, their great-grandparents, have all existed, every day of their lives, in a state of physiological emergency.

They will say things like "Glucose is the body's preferred energy source." (I know because I've heard the claim time and again) Their claim is, of course, ridiculous.

The idea that the energy mode in which man functioned for tens of millions of years is the outlier and glucosis, which has only been the daily mode of operation for a couple thousand years, is the preferred - or 'normal' - mode of fueling the cells is, frankly, absurd.

NOTE: *Many people living in ketosis have, for various reasons, taken a day or two to return to their prior high-carb eating habits. Everyone who has done so reports feeling physically terrible. Some even involuntarily vomit. If glucose was truly the body's preferred energy source, wouldn't the body (which we've determined has an incredible intuitive intelligence) signal a big "Thank you!" by feeling great when the blood glucose level soars? Yet just the opposite occurs. The body reacts precisely as one would expect to the introduction of a toxic substance.*

> *An illusion it will be, so large, so vast it will escape their perception. Those who see it will be thought insane.*
> ~ Anonymous

Further, it is only advances in technology over the last several thousand years that has enabled - permitted - mankind's switch from ketosis to glucosis. The body has no idea what technology man created, and that technology hasn't altered man's hardwired genetic coding one iota.

Until you reached this chapter you likely looked at the the HLS as the body's sole emergency blood glucose response system. I hope you now see that glucosis is the body's emergency blood glucose response system, with the HLS system being merely a component of that broader system.

Why didn't I just come right out and tell you early-on that glucosis is the body's emergency blood glucose response system?

I am the first person in history to state glucosis is the body's emergency blood glucose response system.

Prior to this book glucosis didn't even have a name!

With this book I am publicly declaring glucosis what it really is; a threat to human health.

That is a tremendous paradigm shift encompassing every one of the 8 billion human beings on the planet.

History informs us that humans have a really tough time accepting something as true if it conflicts with everything they've always known.

Foundation has to be laid in order for the human mind to accept a major paradigm shift. Even after laying hundreds of pages of physiological foundation, some readers simply will not be able to accept it.

Whether a new revelation is factual - whether it is the truth - is often not a significant factor in whether it will be accepted by the public.

All truth passes through three stages. First, it is ridiculed. Second, it is violently opposed. Third, it is accepted as being self-evident.
~ Arthur Schopenhauer

By sharing this truth with you now, rather than earlier, I hope I have made it easier for you to see it and accept it.

What of those who say glucose is not toxic to the body? First, I would suggest they read the list of affects it has on the body.

Glucose in its very nature is a toxin to the body, which is why, when in ketosis - as humans were for millions of years - blood glucose is kept at baseline; a safe level for the human body.

In ketosis glucose provides a very limited function; a function that never requires it to increase from baseline.

NOTE1: *Critics of ketosis often refer to it as "the body's starvation mode". That is, of course, nonsensical. That characterization is part of the larger 'scare tactic campaign' employed by the establishment in the hope of keeping you from considering the truth. Millions of people in ketosis are eating 2,500, or 3,500, or even 4,500 calories a day.*

The fastest way to dumb down a population is to scare them.
~ Anonymous

If one reads current medical writings on glucose one would never know cells can/do oxidize fatty acids and ketones for energy. There is no discussion of this incredibly healthy "alternative". (Glucosis is actually the 'alternative'.) The only thing the medical industry tells people is glucose is the body's primary energy source - as if nothing different or better exists. No mention of the fact that it was exactly the "other way around" for all but a few seconds of human history, as the last few thousand years are seen in the full timeline of humankind.

They will write about how confusion occurs when the brain does not get enough glucose (when in glucosis), but make no mention of the fact that virtually everyone who puts their body into ketosis reports an increase in mental clarity and focus they've never before experienced. Yet the brain in running on ketones, not glucose, when the increase in focus and clarity occurs.

In other words, establishment minions are happy to say that lowering carbs (when in glucosis) may result in mental confusion, but are unwilling to say lowering carbs further (in order to be in ketosis) increases mental clarity. Bias? What bias?

> *The last thing a fish would ever notice would be water.*
> *~ Ralph Linton*

Keep in mind that virtually every human being on the planet is placed into glucosis by his/her parents from the time the child begins solid food.

Let me be clear about this; glucosis is not an "automatic" process. It's not just "How it is."

It was done to you! And if you are a parent, you almost certainly did it to your child/children. (That's not a statement of blame or judgment. You didn't know, just as your parents didn't know.)

Because your body was forced into glucosis when you were an infant, and you were taught the glucosis way of eating as you grew, you have always considered glucosis 'normal'. Well…in reality…you didn't 'consider' glucosis at all, just as a fish doesn't notice water.

In nature - in that symbiotic Earth/Man relationship we've discussed - it was exactly the opposite. Man existed in ketosis from birth. Parents could not force the bodies of their children out of ketosis, into glucosis. The food items necessary to do that simply weren't available. Again, the

Earth/Man symbiosis at work.

While 26% of today's kids ages 0 to 19 already have at least one chronic disease, there is no evidence ancient man, which includes their children, had any chronic diseases whatsoever during the millions of years they lived in ketosis.

An intriguing unknown at this time is how ketosian children may fare in terms of modern childhood illnesses. Since ketosis improves immune system function, we can reasonably assume they will fare better than glucosis children. (How much better is yet another reason for N=1 research.)

NOTE2: *Ketosis does not actual improve immune system function. The glucose/insulin cycle degrades immune system performance Ketosis returns the immune system to its proper intended performance level.*

I'm admittedly curious to learn which today's parents value more; social conformity ["My kids need to eat like other kids."] or healthy children who will grow into healthy adults [ketosis].

It is important to note that when a body has been in ketosis for some time glucose is not used by the cells for energy. That would have been the case for our ancient ancestors who were in ketosis from birth. It is also the case for people today who have been in ketosis for some time.

You might reasonably ask why the body maintains a baseline of glucose at all when in ketosis. It does so because there are just a few cells that must have glucose. I won't bore you with the chemistry but they are the red & white blood cells, and the medulla of the kidney. That's it!

NOTE3: *In the absence of carbohydrate intake the body synthesizes just enough glucose to meet these needs, from amino acids preferentially, and fatty acids secondarily.*

NOTE4: *There is disagreement on whether a tiny number of brain cells require glucose or whether all brain cells can function with* ketones. The question is irrelevant to the points being made in this book.

NOTE5: *There are a number of amino acids and fatty acids the body cannot synthesize and must get from dietary intake. These are known as "essential amino acids" and "essential fatty acids" and both are abundant in animal flesh. In other words, the human body is physiologically compelled to eat animal flesh in order to acquire those nutrients. There is no such thing as essential carbs. There is zero physiological need for man to consume any carbohydrates. Ever. Which of course further substantiates ancient man's diet.*

At this point, let's look at a short list of good and bad things that occur from living in glucosis and living in ketosis.

Glucosis (negative):
- Repeated high levels of toxic blood glucose.
- Commensurate high levels of harmful insulin.
- Daily abuse of the emergency Hepatic Lipid System
- Insulin Resistance
- Syndrome X
- Alzheimers
- Heart Disease
- Type-2 Diabetes
- Cancer
- Obesity
- And more…

Glucosis (positive):
- Nothing. Physiologically speaking, quite literally nothing.

Ketosis (negative):
- Scientific research (to the very limited extent performed), and the

personal experience of millions of people living in ketosis, has shown zero negatives.

Ketosis (positive):
- Experience none of the things in the "Glucose (negative)" list (above).
- A higher degree of mental clarity and focus.
- More abundant energy.
- Not merely avoiding obesity, but effortlessly dropping all excess body fat.
- Blood glucose maintained at the healthy non-toxic baseline.
- Insulin maintained at non-harmful baseline.
- Increase in efficiency of the body's immune system.
- Avoidance of virtually all chronic disease.
- And more...

The establishment has expressed many "concerns" about Americans living in ketosis.

Because millions of people have been in ketosis for several months at a time with zero negative affects, the establishment has shifted it's faux concern to being in ketosis "long-term".

Of course the establishment conveniently fails to note there are, at a minimum, hundreds of thousands of people in the U.S. (likely millions worldwide) who have lived in ketosis for years without a single claim by anyone of any health problem attributable to ketosis.

> *One's destination is never a place but*
> *rather a new way of looking at things.*
> *~ Henry Miller*

The establishment has even gone so far as to say it is concerned that by living in ketosis people are adopting a state of existence that is in conflict with our genetic coding! (Now you see why it is important for various groups/persons to lie about the diet of ancient man.)

People living in ketosis for years are the healthiest people in America. And **that** is precisely what concerns the establishment.

> To keep this book short, the subject of how to place your body into ketosis, and continue "living in ketosis", is not addressed herein. If you'd like to receive (at no cost) a paper I wrote entitled "Getting Started In Ketosis", simply go to drreality.news/reports and input your email address. The document will automatically be sent to you. (Your contact information will always remain confidential.)

CHAPTER THIRTEEN

A BRAVE NEW WORLD

"He who makes no enemies makes no difference."
~ J.W. Barlament

From where might we expect opposition to this message?

We've all heard the wise old adage, "Follow the money", so let's start with the economics of the subject.

As odd as it may sound to you on first blush, much of the American economy has its foundation anchored in glucosis. Trillions of dollars each year depend on preserving the lie that glucosis is natural, proper, and healthy.

The U.S. Gross Domestic Product for 2018 (as I write this, the latest year for which we have a solid figure) was $21.3 trillion. The medical and food industries, combined, account for 23% of that. In other words, those two industries represent $4.9 trillion in revenue.

What would happen to those industries if, for the sake of illustration, all 330,000,000 Americans understood the truth and acted on it tomorrow, choosing to live in ketosis. Answer: Those industries would be devastated, at best. "Demolished" might be more accurate.

Let's start with the medical industry and then move on to the food industry.

To accurately detail the impact on the medical industry of all Americans living in ketosis would require a large team of experts and a lengthy report. Nevertheless, there are ramifications that don't require an in-depth study to understand.

First, consider just the single aspect of chronic disease dropping to almost nothing.

Second, consider that when in ketosis - because the cells are no longer being continuously poisoned - the body experiences healthfulness in virtually every area for which the medical industry now employs specialists.

Again, without a detailed report we cannot know precise numbers, but let's use some reasonable guesstimates to visualize the economic consequences.

Let's say 3 years after all Americans were living in ketosis the need for medical care was reduced by 80%. (A number I believe to be fairly accurate). That means the current $3.8 trillion in revenue generated by the medical industry would be reduced by a staggering $3 trillion! Total revenues would drop from $3.8 trillion to $800 billion.

What do you think the medical industry is willing to do to protect that $3 trillion from being lost? (Hint: Pretty much the same thing they did to create it!)

> *"When morality comes up against profit,*
> *it is seldom that profit loses."*
> *~ Shirley Chisholm*

The food industry is more difficult to analyze casually because it involves food you take home as well as food you consume outside the home. Nevertheless, here are some obvious ramifications for that industry.

One consequence is immediately apparent. The processed food industry (at least as it exists today) would cease to exist.

The restaurant industry would experience massive disruption. Those that could quickly and creatively adjust their menus would survive and thrive. Most would go out of business, being replaced over time by ketosis-based restaurants.

The fast food segment as we know it would cease to exist. However, because living in ketosis would not change the propensity of Americans to "eat on the run", we would soon see a reinvented fast food industry, offering what people living in ketosis want and are willing to buy.

If all Americans decided to get healthy by living in ketosis the public would no longer have any interest in buying most of what is currently on grocery store shelves (which exists to fill us with carbohydrates).

While I guesstimate purchases of the grocery items we see on the shelves today would drop by 90%, that would be off-set by a corresponding increase in the sales of foods people in ketosis eat.

As an illustrative example, a person who may have consumed 3,000 calories a day in glucosis may also consume 3,000 calories a day in ketosis. In other words, people would still purchase food items providing the same or similar volume of necessary calories. They would simply purchase different food items from a more narrow list. (Grocery stores could likely cut their square footage by 70%.)

Much like the restaurant industry, grocery industry giants that could adapt would survive. Many would not. But new operations would rise to take their place.

When we speak of the grocery and restaurant industries "adapting", that requires a look at the sources from which they buy their goods. Providers of foods that keep people in ketosis would flourish. (A significant understatement.) Their #1 problem in the short term would be meeting the increased need. That would be resolved quickly enough.

Then there is the impact on peripheral industries.

I imagine the segment of supplement industry that produces things like sports drinks, protein bars (virtually always packed full of carbs), powders to (supposedly) help people gain muscle, pre-workout drinks, etc., is nimble enough to adapt quickly. Instead of bars, powders, and drinks high in carbs, they would quickly convert to products containing the necessary macro ratios to support ketosis.

The vitamin (and other "pills & powders") segment would likely see a steep decline in sales. Healthy people have much less need/desire of such things.

Except in rare cases of genetically-induced issues, there will be little use for jobs such as "nutritionist" and "dietitian".

I don't want to avoid stating the obvious. The result of every American living in ketosis would be an economic upheaval of massive proportion. I do not mean to minimize that reality by also stating that it would be, in reality, an economic realignment. With the possible exception of the processed food industry, every other segment would simply reinvent themselves to meet the new customer buying criteria.

And America would be healthy again.

But not to worry; that is not going to happen.

Most Americans will choose to continue in glucosis, so the outcome for these industries is nowhere near as dire as characterized above.

That said, these industries will spend whatever amount of money it takes to fight against the truth of the glucosis/ketosis issue for 2 reasons:

1) Even a modest reduction in sales represents billions in lost revenue.
2) They are terrified this truth will gain traction with the public.

We know that when the public chooses a healthier way of eating the food industry experiences decreased revenue.

The Atkins Diet was very popular in the U.S. in 2003 and 2004 and promoted a significant reduction in carb intake. It is estimated that in 2003 the popularity of Adkins resulted in an 8.2% drop in pasta sales and a 4.6% decrease in rice sales. If the facts revealed in this book gain widespread acceptance, the market impact would be far greater than what occurred in 2003.

It should go without saying that the multi-trillion dollar industries at risk won't just fund media attacks against the message I've shared with you here. They will also initiate attacks on the messenger. Fun times ahead!

Before I leave the industry side of the discussion, you may recall in chapter 7 I mentioned society finds itself in this predicament due to misuse of various emerging technologies over the last 12,000 years; especially the last 150 years. I also stated technology can be used to great benefit or harm. The good news is, if we - the consumers - demand change, current technology can be of tremendous value in creating the change we desire.

> *"I was brought up to believe that the only thing worth doing was to add to the sum of accurate information in the world."*
> ~ Margaret Mead

A good question at this point is, "How much do the titans of the food and medical industries know about the things we've discussed?"

That's uncertain, but allow me to share a couple of thoughts with you.

Though I doubt they currently understand that glucosis is the body's emergency response system to eliminate blood glucose above baseline, I'm 100% certain they know that chronic disease is driven by the glucose/insulin cycle forced upon the body by high carb intake.

My certainty that they know is based on the fact that there is a complete blackout on the funding of any research that might even remotely hint at the fact that the glucose/insulin cycle drives chronic disease. The funding blackout is so complete there can be no question it is intentional.

There have been a number of nutritional and medical research papers in which the authors state without equivocation the numerous positive outcomes of small test groups existing in ketosis for short periods of time (usually measured in weeks). These remarks appear in the "discussion" portion of the reports, which does not constitute the research "findings".

Yet it seems that in order not to cross any line from which their careers cannot recover, those mentions always conclude with some wishy-washy sentiment suggesting that...perhaps...maybe...someday...we should consider thinking about looking at this.

Despite these mentions in research papers, not one penny has been provided for something as a simple as an observational study comparing the health of those living in glucosis with those living in ketosis. With the existence of mobile apps, and automated in-the-background data transfer, not only would it be a breeze, but inexpensive.

How about a modest size "proof of concept" first effort; say...200 participants? 100? 60? 40?

Hell, Dave Feldman has thousands of people already participating with his effort - and that is completely voluntary, driven by his YouTube and Twitter presence, and his website.

> *A lot of things were thought to be crazy before they succeeded*
> *~ Anonymous*

Yet somehow we are to believe that the multi-billion dollar nutritional research industry can't find a way to put together a simple and inexpensive observational study on what is clearly the most important health question of our time? Oh wait...I know...it's merely an oversight!

You may also be wondering why, if the successes of ketosis are touted in various research papers why you haven't heard this from your own doctor. Isn't that the $64,000 question!

The answer is multifaceted.

First, after leaving medical school the vast majority of doctors no longer read research studies. If they read any at all, they read only the abstract and/or summary; never the body of the report.

Second, while the world of science and information is a big place, the lens through which doctors see information is very narrow. If you can envision a doctor taking two drinking straws, one in each hand, raising them to his/her eyes and using them to see through like binoculars, that is a fairly accurate analogy of how constricted is their willingness to view/consider data.

Why is their willingness to look at data so constricted?

Let me share a real-world analogy with you.

My first book, "Income Tax: Shattering The Myths", has been read by countless thousands of Americans. Do you know who does not read it? Accountants, CPAs, and tax attorneys. Why? Because it would present a painful conundrum for them.

The 408 pages of history, constitutional analysis, statutes, regulations, U.S. Supreme Court decisions, Treasury Decisions, Treasury Orders, and a myriad of other internal government resources, laid out and connected in perfect clarity, do not permit any conclusion other than the truth. Yet the truth differs from the non-factual 'belief system' demanded of accountants, CPAs, and tax attorneys in order to do their job as directed by their industry masters.

> *It is difficult to get a man to understand something
> when his salary depends on his not understanding it.*
> ~ Upton Sinclair

The same is true of the vast majority of doctors. If the facts conflict with what the medical establishment - the medical industry power structure - requires doctors to 'believe' and regurgitate on demand, just like accountants, CPAs, and tax attorneys, they will squeeze their eyes tightly shut against the truth and do as their industry masters dictate.

> "The desire for safety stands against
> every great and noble enterprise."
> ~ Tacitus

Further, doctors fear that if they give you 'medical advice' that is not 100% sanctioned by their industry's leadership, they will be sued. I admit to having grown weary of hearing medical personnel tell me doctors feel compelled to share medical industry dogma with their patients out of fear of being sued.

Fortunately you and I are not constrained by what establishment minions think, say, or believe.

Exploring the correlation between what we put in our mouths and what happens to our bodies requires nothing more than a desire to investigate the truth - and perhaps a dash of adventurism.

In a study performed in 1975, Dr. Joseph R. Kraft demonstrated (via a very specific and previously unperformed manner of testing) that 75% of Americans tested had abnormal insulin response patterns and had, as Dr. Kraft coined it - 'diabetes in-situ' - meaning although conventional testing would not reveal it (and still does not), these people were on their way to becoming diabetics. (Dr. Kraft's results were confirmed in

2013 by another study that produced a second set of troubling insulin response patterns know as "Hayashi patterns".)

Kraft's finding that 75% of all Americans are on their way to type-2 diabetes was in 1975! The public's diet has gotten considerable worse since then. Can we even guess what the percentage might be today? 90% seems a shoe-in. Perhaps higher.

Let us say for the sake of argument the current number is 93%. Who is in the 93% and who is in the 7%? Should I point out that 93% of all of you reading these words are - whether you realize it or not - heading inexorably toward type-2 diabetes? Well...'inexorably' if you continue in glucosis.

It is important to note that looking good and feeling good is completely unrelated to Kraft's findings. His specialized testing protocols revealed abnormal insulin response patterns in people years before their situation progressed to weight-gain and the presentation of disease.

NOTE1: *Vegetarianism, a plant-based diet, and/or the diet of those in certain Asian countries are sometimes raised as a refutation of what you've read here. None is a refutation. They merely constitute a failure to correctly understand the pertinent physiological issues involved.*

Kraft's specialized protocol reveals that living in glucosis is harming you years before the harm is discernible by you.

I hope there is no question in your mind at this point how to stop 'diabetes in-situ' - and diabetes generally - dead in its tracks and get healthy.

At this point one might wonder why the medical industry, primarily through the media and 'medical advice', still directs Americans to lower dietary fat - sat-fat in particular - to prevent heart disease, when doing so has no bearing on heart disease.

Simple. Let's say a person's energy needs are met by consuming 2,500 calories a day. There are only 3 macros from which to obtain those calories; protein, carbs, and fat. If people feel free to consume more fat, they will likely reduce their consumption of carbohydrate intensive foods. If they increase calories from fat and decrease calories from carbs, they may accidentally discover something - such as they start losing weight and their blood test numbers improve!

Remember, the medical industry stands to lose trillions in revenue, and jobs, if Americans get healthy.

Can you imagine how little an office visit to a neighborhood doctor would cost if everyone was healthy and didn't have much need for doctors anymore?

There are few other readily anticipated adversaries to the message in this book.

There are people who believe animal-based food production must stop because they assert it contributes to global warming. There are also people who feel eating animals is cruel.

My suggestion is to simply eat the way you need to eat to be healthy, and let them eat their way. When enough people adopt ketosis, with results readily apparent, the debate - the one that hasn't even started yet - will be over.

If you are not your own doctor, you are a fool.
~ Hippocrates

Let's talk about N=1.

At the beginning of this book I told you I do not want you to believe me, and I encouraged you to do your own research.

Research comes in different forms.

You can do as I have done and spend untold hours buried in studies, scholarly papers, research reports on-line, endless internet searches for obscure facts you'll feel the need to run down, etc. And that is a great way to learn!

I also did N=1. I performed the intellectual side and the practical application side.

The intellectual side was easy in the sense that it was not "work" for me. I love every moment of the learning experience. Yet I am aware not everyone is like me.

So...what if you just want to cut to the chase and do the N=1 part of the research to acid test what you've learned in this book; perhaps to determine whether what I've shared with you is factual?

If you have reached this point and do not accept what you've read, then I **really** want you to do N=1!

Go get a lipid panel blood test and then put yourself in ketosis for 4 months. Then go get another lipid panel. Between how you feel and what your blood test results show you, you will become a believer!

NOTE2: *LDL-C sometimes rises when in ketosis. As mentioned earlier, there is nothing bad, wrong, or harmful about your body setting a cholesterol level higher than establishment minions tell you it should be. (The exception being if accompanied by high triglycerides.)*

What are the risks of putting yourself in ketosis? None. Zero. Zilch. Nada.

(If you are a type-1 diabetic you must continue responsible monitoring of your blood glucose and use of insulin as indicated.)

What are the benefits? You've spent 13 chapters reading them.

If I were to summarize the benefit for you in a single sentence, it would be...

Living in ketosis is the means of being as healthy as a human can be while reducing your odds of getting modern chronic disease to virtually nil.

Hopefully you've gotten the message that this entire subject revolves around the choices we make, individually, and collectively (as a nation - and the world).

> *"Never doubt that a small group of thoughtful,*
> *committed, citizens can change the world.*
> *Indeed, it is the only thing that ever has."*
> *~ Margaret Mead*

It is my hope this book encourages so many people to make themselves healthy again by putting themselves in ketosis that America winds up with an 'army' of amazingly healthy Ketosians. The establishment's lies would collapse under the weight of that kind of evidence; the kind everyone can see for themselves when by looking at their co-workers, chatting with neighbors, or listening to family members.

> *Be ashamed to die until you have won some victory for humanity*
> *~ Horace Mann*

As you consider where you'd like to be, let's set some expectations about going into ketosis.

Will you notice the changes immediately. No.

NOTE3: *Most people experience 2-4 days of not feeling great as their body transitions from glucosis to ketosis. This is because the cells are expecting glucose you are no longer giving them. This brief period is sometimes called "Keto flu" (though it's not remotely like having the flu). When you're body flips into ketosis (usually by day-4) you will immediately feel great.*

Within a few weeks? Absolutely.

Previous (pre-ketosis) problematic blood test numbers should look considerably improved by the 60-day mark, with continued improvement for perhaps another 60 days. In other words, for the average person who chooses ketosis, blood work should be completely healthy (or damn close to it) within 120 days. (This assumes you commit 100% and are not merely playing at it, such as eating one meal geared to ketosis but another geared to glucose. This also speaks to the average person in ketosis. A person who is, as an example, 150 pounds overweight, will obviously have a longer timeline.)

If you've lived in glucosis for 40, 50, or 60 years, the estimate is that a complete internal transformation takes roughly 20 to 24 months. Bio-individuality applies.

With that said, will living in ketosis repair 100% of the damage done by the long-term affects of glucosis? I imagine we'd all like to think so, but the reality is we don't know.

While the real-life experience of millions of people living in ketosis is that it produces an astounding reversal in virtually every aspect of poor health, there hasn't been any scientific research performed that would tell us whether, over time, ketosis can/will repair every bit of damage done to the body by glucosis. In fact, because the establishment considers glucosis normal and healthy (or at least wants you to believe that) there have been no studies to determine the extent of harm glucosis does to the body, including damage that may not be apparent.

You may have noticed I've not spent a lot of time discussing the loss of body fat that results from living in ketosis. There are a couple of reasons for that.

First, I wanted to ensure this book is not mischaracterized as a "diet" or "weight loss" book. (There are more than enough books out there merely re-arranging chairs on the deck of the Titanic.)

Second, it is important to recognize the fundamental benefits of living in ketosis is not about "looking good". Looking good isn't the goal, in my opinion. Looking good is a consequence of living properly and being healthy!

Phrased another way, just as the accumulation of excess body fat is a visual sign a person is mistreating his/her body, so the absence of excess body fat when living in ketosis is the visual sign you're living right.

> **NOTE4:** *A person may not end up as lean as he/she wishes to be. But in ketosis you will end up as lean as your genetic coding permits. Bio-individuality.*

At the end of the Introduction I expressed a desire that when you arrived at this moment you would say, "That was an astounding experience. **This changes everything.**" I hope that is how you feel about what you discovered in this book.

Americans have been scammed by their government, the nutritional research industry, and the medical industry for at least 50 years. When you add up all the ugly facts it's hard to see the con-job any way other than profits have been - and continue to be - what motivates them to lie to you.

If your health must be sacrificed upon the alter of their trillion dollar revenues in order to ensure the wealth keeps rolling in, that's a sacrifice they're willing to make - and always have been.

Much of what appears in this book is precisely the information the subterfuge is intended to ensure you never discover.

Over the last 5 decades America has slipped evermore deeply into the worst health crisis the world has ever seen. In that **half century** the one thing that has **never** even been mentioned, in the full dynamic I have described for you here, is the epic life-altering distinction between glucosis and ketosis.

The time is now.

To keep this book short, the subject of how to place your body into ketosis, and continue "living in ketosis", is not addressed herein. If you'd like to receive (at no cost) a paper I wrote entitled "Getting Started In Ketosis", simply go to drreality.news/reports and input your email address. The document will automatically be sent to you. (Your contact information will always remain confidential.)

INSULIN INTOLERANCE HYPOTHESIS

Type-1 diabetes is the result of a genetic abnormality with no known cure.

Complicating matters is that science has not been able to isolate the precise gene mutation(s) that cause type-1 diabetes, so no one knows who has the mutation(s) that invoke type-1 - until the disease strikes.

What you are reading is my hypothesis concerning a means of preventing the onset of type-1 diabetes. This hypothesis is best considered in the context of the revelations about physiology I make in *Body Science*.

There is no existing scientific data supporting my hypothesis.

The reason there is no scientific data to support my hypothesis is because in the U.S. there is a complete blackout on any and all research the result of which may implicate the daily abuse of the HLS as being in any way problematic to your heath.

Since probably 99.8% of humans on Earth exist in glucosis - and have for at least several hundred years - it stands to reason that everyone born during those years with the genetic mutation for type-1 diabetes have lived their lives in glucosis.

We know that type-1 is often passed on genetically. In other words, if one or more of your parents, grandparents, or great grandparents had type-1, you have significantly greater odds of having the type-1 diabetes gene mutation(s).

From a functional perspective, at its core, type-1 diabetes is the result of an improper autoimmune response by the body. That improper response causes the body's autoimmune system to attack the beta cells in the pancreas, which are the insulin producing cells. The autoimmune attack destroys the beta cells, thus greatly reducing, or ending, the body's ability to produce insulin.

We know that the rates of people experiencing the onset of type-1 diabetes varies dramatically from location-to-location within the same basic zone, such as Europe where there is up to a 10-fold difference in the rate from one locale to another. This strongly suggests there are factors beyond genetics that determine who experiences onset and who does not.

Further, the percentage of people with type-1 diabetes, while still a tiny fraction of all people diagnosed with diabetes, has increased dramatically in the last 150 years. Since a dramatic increase in such a short period of time cannot be the result of an addition or expansion of genetic issues, it is virtual certainly that environmental and/or behavioral factors are playing a significant role.

Interestingly, there are a number autoantibodies involved in type-1 diabetes, but the detection of these autoantibodies in advance of onset does not necessarily mean the person will get type-1 diabetes.

I mentioned the "onset" of type-1.

Because there is no known way to prevent the gene mutation that causes type-1, nor the production of the various autoantibodies which often, but not always, indicate onset is likely, if one does not wish to live with type-1, the only remaining avenue of consideration is preventing or forestalling its onset - if such a thing is possible.

It is to that possibility my hypothesis speaks.

Because of the factors mentioned, there may be a course of action to prevent its onset.

As mentioned, virtually everyone alive today who has type-1 was born into a glucosis world. A child's first baby food assures he/she will exist in glucosis. From the time they eat small bites of solid food, they are fed high-carb items. Having cut their teeth - literally - on high-carb food, that is what they prefer. And so the pattern is set.

I hope one of the big take-away messages from *Body Science* is how amazing - how intelligent - our bodies are.

In today's society there is essentially zero awareness of how injurious high insulin is to our health. We have no awareness of this because the powers who control the research money won't allow any research to be funded that would validate what hopefully you understand at this point.

Let's consider that the body has an innate knowledge that repeated high insulin is destructive. Yet it bears that burden day-in & day-out, for decades, until the systems impacted begin to deteriorate from the constant abuse. The first signs of that deterioration being things like insulin resistance and Syndrome X.

Now let us say that some bodies - with genetic mutations - are dramatically more sensitive to the threat posed by high insulin. From the first time insulin spikes, in those genetically mutated bodies an inner silent alarm starts shrieking. If those bodies could speak independent of the conscious entities inhabiting them, they would be screaming, "STOP! I CANNOT COPE WITH INSULIN SPIKES. IT IS HARMFUL AND EVENTUALLY POTENTIALLY DEADLY!"

The body has no ability to communicate that message.

But perhaps, if it is ultra-hyper-sensitive to the threat, it will act to protect itself from toxic high insulin spikes by making sure they can no longer occur; by sending autoantibodies to destroy the source of the insulin - the pancreatic beta cells.

In this scenario, while the body is clearly "broken" in the sense that it goes off the rails in its attempt to save itself, its attempt at survival may give us a path by which to prevent the onset of type-1 (if there is validity to my hypothesis).

That path is for a child to remain in ketosis from birth. In ketosis there will never be any high insulin. Moderate insulin will occur if considerable protein is consumed at one sitting, but no high insulin, ever.

Here is the way I have characterized this hypothesis to my friends in casual conversation.

Imagine you're locked in a room with a psychotic killer. He has lots of sharp deadly knives strapped to his body and you have nothing with which to defend yourself. If he wants to take your life there is little you can do to stop him.

A gentleman who spent years studying psychology, and this particular psycho killer, told you the psycho only wants to harm people who play music loud. The kind of music is irrelevant, it's the volume that's sets him off.

The psycho killer just sits in the corner looking down. He doesn't seems to notice you.

You have a stereo in the room and you turn it on to a low volume. Psycho killer continues minding his own business. You turn it up a wee bit more. No reaction. You turn it up a bit more and you see his eyebrow move slightly and his eyes dart quickly over to your side of the room.

How much more would you want to turn up the stereo? By the time you reach a volume that sends him into a psychotic rage, and he murders you, it's too late to go back and do it differently. You pushed it and lost.

The psycho killer just sitting in the corner is analogous to the body being poised to launch the first autoantibodies, thus signaling the attack against the beta cells is likely impending.

Turning up the volume on the music is parallel to "turning up the volume" on insulin levels; setting the killer off.

Your murder is analogous to the onset of type-1 diabetes. (Prior to the discovery of insulin in 1921, type-1 diabetes was a death sentence.)

All one need do to not be murdered by the psycho killer is refrain from playing music loud.

Likewise, if my theory is valid, all one need do to avoid type-1 onset is never spike insulin.

Has this theory every been tested? No, of course not. It's too straightforward. If successful it would likely be considered too revealing of the dangers of America's carb-intensive - hence insulin-intensive - diet.

My hypothesis is not too terribly far from one of the hypotheses offered by the mainstream diabetic medical industry, known as the "accelerator hypothesis."

Interestingly, when my study of physiology led me to visualize the hypothesis you are reading, I did not know of the existence of the "accelerator hypothesis."

> "...in physiology or medicine, it's about making discoveries, and you don't have to be clever to make a discovery, I don't think; it just comes up and punches you on the nose."
> ~ Tim Hunt

The "accelerator hypothesis" suggests that over-nutrition, or over-feeding, in childhood leads to insulin resistance, raising the risk of diabetes. Yet we know that in order for a normal body to develop insulin resistance it must experience repeated high insulin levels over a lengthy period, usually measured in decades.

The "Insulin Intolerance Hypothesis" is distinct from the "accelerator hypothesis" in 3 specific characteristics.

1. It tightens the focus from generalities to high insulin levels specifically.
2. It posits that the body of a person with the type-1 enabling genetic mutation(s) perceives repeated high insulin as a "clear and present danger", reacting improperly to terminate the "threat".
3. Suggests the very specific action of living in ketosis as a means of preventing type-1 onset.

Considering there is no way to know who will get type-1 and who will not, how would parents determine if their child should be kept in ketosis? Certainly this would be advised for a child with multiple relatives who have (or had) type-1 because the odds would be high for that child getting type-1.

How about if one parent has type-1?

Did I mention there is no danger whatsoever from placing a child in ketosis from his/her first non-breastmilk food? Since there is nothing dangerous in doing so - after all, that's how human children existed for millions of years - if there is any reason to suspect type-1 is a genetic possibility for your child, why not be safe and start his/her life in ketosis?

How would this theory be proven correct?

First, let me say that if it prevents people from getting type-1 I'm not overly concerned about it being "proven" by a corrupt money-hungry medical establishment. The only benefit to it being proven correct is it would receive greater public acceptance, and thus improve more lives.

I'm not holding my breath for the establishment to test the theory...ever.... even if 100% of those living in ketosis from birth never get type-1 diabetes.

In fact, if that happened, I would expect it to harden the establishment's position to never officially test it.

Since living in ketosis is the healthiest way to live for all human beings, we're talking about a choice that may prevent the onset of type-1 and is incredibly healthy at the same time.

Seems like a win-win to me!

ACKNOWLEDGEMENTS

Ernie Fowlke is a long-time friend. Yet "friend" may not be an adequate word. There are few people who value truth above all else. I find inestimable value in such people. Ernie is one such man. Over the years Ernie has been a consistent supporter of my work in ways large and small. He has taken his valuable time (yet again) to assist me in this project. I do not know what I have done to earn his friendship and support, but it is an understatement to say I am grateful for it.

Robert F. Morrow is always willing to lend his significant talents to an endeavor that is "a cause greater than one's self." If Robert sees a positive impact can/will result for millions of people, he is ready and willing to come along side and help. I am grateful to Robert, not just for his assistance, but for his friendship, as well as for his easy and gracious manner. If one questions the meaning of the word "gentleman", one need only spend 10 minutes with Robert to see the word exemplified.

My appreciation to **Tyler Schmeling** for the rear cover photo. Not only is Tyler an incredibly talented photographer, he is a Physician's Assistant, holding a Master's Degree in Nursing, and has an intellect second to none. He is an amazing friend.

Scott Settles has been a friend for many years, often expressing appreciation for the volume of material I have put into the pubic domain over the decades (at no cost), simply for the benefit of my fellow man. When changes to back-of-the-house technology were needed to accompany the introduction of Body Science, Scott immediately offered to bring his substantial skills and experience to the table. Men like Scott are rare in this life.

Test-Reader: I wish to thank the **test readers**. These involved and engaged people - with jobs, children, and other demands - carved out the time to read the manuscript version and share their views with me. Test readers ran the gamut, including doctors, corporate executives, stay-at-home moms, fitness professionals, retirees, entrepreneurs, scientists, medical students, and business owners. All of them read the manuscript "cold", i.e. with no prior knowledge of the revelations they would encounter. The #1 thing I asked was to share their thoughts concerning content & flow with an eye to making the message - the Big Picture - as easy for readers to comprehend as the material permits. If this book is clear and flows well, their input was an important contribution to that.

Those Who Came Before - There are the countless private researchers (which in this context includes some journalists) whose curiosity, concern, and diligence resulted in public access/availability to many of the facts you find in this book. Without them this book would not have been possible.